Raising Healthy Rabbits

by

Dr. W. Sheldon Bivin **Dr. William W. King**

Division of Laboratory Animal Medicine
Louisiana State University
School of Veterinary Medicine
South Stadium Drive
Baton Rouge, Louisiana 70803

Christian Veterinary Mission

Christian Veterinary Mission

Current Book Information at:
www.cvm.org

Permission requests should be addressed to: vetbooks@cvm.org

OR

Christian Veterinary Mission
www.cvm.org

THIS IS A REVISED THIRD EDITION

It can be improved with your help. If you are a veterinarian, veterinary technician, animal husbandry worker, missionary, development worker or anyone with ideas or suggestions for ways this booklet could be changed to better meet the needs of your people, please write to the authors at Christian Veterinary Mission.

Thank you for your help.

Raising Healthy Animals Series

Every year, thousands of people around the world struggle to survive because they don't have the right knowledge, skills and resources to care for their animals. Christian Veterinary Mission (CVM) sends veterinary professionals to live and work alongside many of these people to encourage them and provide them with not only much needed veterinary expertise, but also the hope that is only found in Christ. CVM veterinarians build lasting relationships with individuals and communities, helping them be transformed through Christ's love.

CVM, in its effort to be meaningfully involved in work in the developing world, quickly found there was little appropriate educational material available. CVM set about developing basic resource materials in animal husbandry for farmers and agricultural workers. Apparently, they met a real need, as these books have been accepted throughout the developing nations of the world.

The series of books published by Christian Veterinary Mission includes the following in order of publication:

Raising Healthy Pigs *	Drugs and Their Usage
Raising Healthy Rabbits *	Where There Is No Animal Doctor
Raising Healthy Fish	Raising Healthy Horses
Raising Healthy Cattle	Zoonoses: Animal Diseases That Affect Humans
Raising Healthy Poultry *+	Raising Healthy Honey Bees
Raising Healthy Goats *	Slaughter and Preservation of Meat
Raising Healthy Sheep	Disease and Parasite Prevention in Farm Animals

[Also available in: * Spanish + French].

CVM fieldworkers have also developed specific training materials for the countries in which they work.

All of these books have been put together by Christian men and women; in a labor of love and service, for people in need throughout the world. It demonstrates dedication to their profession, service to humanity and a witness to their faith. We hope that they are a help to you in developing an appropriate livestock program to meet your needs. We pray God's blessing on their use.

Leroy Dorminy
CVM Founder

TABLE OF CONTENTS

SECTION VI—Diseases

SECTION VII—Sanitation and Slaughtering

SECTION VIII—Rabbit Production in Various Parts of the World

Section 1

Acknowledgements
Introduction

ACKNOWLEDGEMENTS

The authors wish to thank the Christian Veterinary Mission for taking the initiative to seek authorship and publish this type of material. We sincerely believe that this information will allow many people to provide a cheap source of high quality protein, and we are very proud to have been asked to participate. Obviously, the information for such a text comes from any sources; however, the authors wish to express special acknowledgements and thanks to two organizations: The Oregon State Rabbit Research Center, Oregon State University, Corvallis, Oregon, 97331; and the Agricultural Research Service, U.S. Department of Agriculture.

The authors also wish to express deep appreciation to these secretaries, Dawn Best and Beth Wilson, for their assistance in editing and typing this text.

INTRODUCTION

Many developing countries have populations which are suffering from protein deficiencies. As a means of alleviating the need for protein, government agricultural research centers and Christian Veterinary Missions are encouraging the production of fish, poultry, swine, some rodents such as capybara and guinea pigs, and rabbits by small family units or villages.

This text will outline essential information required for rabbit production. There are several specific advantages for raising rabbits. They do not compete for grains used for human consumption, since they can reproduce and grow on low-grain, high-forage diets. Unlike chickens, they can be successfully raised on a diet consisting largely of forage, plants, grasses and shrubs. Because they are small animals, these food substances can be easily collected by a small family or village unit. Another advantage is that rabbits can use these sources of protein more efficiently than most other animals because they have a high feed conversion ratio. Weanling rabbits can gain one kilogram of body weight for each 2.5-4.0 kilograms of plants they eat. For beef cattle on a similar diet, 12-15 kilograms of feed are required per kilogram of body weight. This means that a hectare of alfalfa can produce at least five times more rabbit meat as compared to cattle raised under similar circumstances. In many areas of the world the keeping of large animals for meat is not feasible. This is primarily due to their large feed and space requirements, the long time taken to reach slaughter weight, and the problem of using a large amount of meat at one time when refrigeration and electricity are lacking. In these instances, the rabbit has the distinct advantage of serving as a "biological refrigerator", since the meat from one animal can be quickly consumed without the need for storage. Another advantage of the rabbit is its high reproductive rate and its ability to thrive on weeds, tropical forages, vegetable leaves and tops, and table scraps. Twelve good does, producing five litters each year, and raising seven fryers to the litter, will produce 420 fryers per year for the table. At a 50% average dress out, they will produce between 380 and 475 kilograms of high quality white meat, which is enough to satisfy the needs of an entire village for a year.

Because there is a pressing need for protein in millions of people living in predominantly rural environments where forages are plentiful, Christian Veterinary Missions (CVM) highly recommends following the outlines presented in this text regarding raising healthy rabbits for food. As competition for grain intensifies between people and animals, it is possible that rabbits will assume an increasingly important role as a source of food.

This manual is designed to give basic information that has been shown to be successful in a variety of situations. We have often opted to describe methods that may be too sophisticated or expensive to apply in every case in an effort to illustrate the optimal condition. Obviously, not everyone will be able to install stainless steel caging and automatic watering systems, but to omit these would impede individuals aspiring to achieve the ideal. In most cases we must rely on your ingenuity and creativity to "make do" for your particular climate and cost restraints.

Also note that the majority of information available in rabbit production relates to the European or domestic rabbit, Oryctolagus cuniculi. More specifically, most production units have used the New Zealand white breed.Most of the information in this text is therefore based on the domestic rabbit.There are native species and breeds of rabbits in various parts of the world. Keeping these, as well as cross-breeding with domestic species, has been successful in many countries. We have attempted, whenever possible, to include information of other species or breeds that have been reported to function well in a production setting.

Section 2

Housing, Equipment And Furniture

HOUSING, EQUIPMENT AND FURNITURE

Permanent rabbit caging should be constructed before the arrival of the first rabbits. This avoids the situation where temporary housing becomes, by default, the permanent housing for the rabbitry.

The site of the rabbit hutches should have fresh air circulation and good drainage. Because these are burrowing animals, the flooring must be durable. If solid floors are used, bedding must be used to absorb excrement.Outdoor hutches can be the best means of meeting the fresh air requirement and, if located properly, even dirt floors can be adequately drained.Outdoor suspended wire hutches with free air flow and dirt under the wire mesh cage floor also cuts down on fly problems.

EQUIPMENT

Rabbits are housed in hutches made of a variety of material. A simple house can be built using a wood frame with a solid wood roof and floor. The sides of the cage may be covered with a chicken wire mesh and the floor can be covered with hay for bedding. Because they produce a lot of feces and urine, the flooring should allow these materials to easily pass through. If welded wire is available in a fine mesh, this is preferred over wood as a flooring material (Figure 1 and 2).

The rabbit building, which is required for larger operations, can be opened sided in warmer climates. The side toward the prevailing winds should be closed. Even in areas with some cool or cold weather, a one sided building with curtains made of plastic or some other material can be used. In a completely enclosed building, a ventilating system delivering 14 to 20 air changes per hour is required and can defeat the economics of rabbit raising. In either case, abundant ventilation is of tremendous value in controlling diseases such as snuffles (see disease section).

In many areas of the world, bamboo can be used to construct the flooring, sides and roof on a wood frame rabbit hutch (Figure 5). Although not as effective as wire, split bamboo can be used to form slatted floors and walls. The smooth, rounded sides of split bamboo should be placed on the inside of the hutch to prevent chewing and to facilitate cleaning and sanitation.It is also important to place the split bamboo slats close enough together to prevent injury to the feet. Usually a 1-2 cm spacing is adequate.

A more permanent hutch can be made with a wooden frame and wire mesh (Figure 6). Commercial rabbitries often use hanging, or quonset-style cages in an open sided building with a solid gravel, dirt or concrete floor (Figure 3). Cages should be constructed of 14-16 gauge galvanized or rubber coated wire. The sides and top can be 2.5 cm x 5.0 cm mesh and the floor wire should be 1 cm x 2 cm. The average height for all cages is 51 cm; whereas, the floor area should be 75 cm x

75 cm if the young are removed at four weeks of age, and 75 cm x 91 cm if the young are removed at six weeks of age. Young does, prior to their first breeding, can be kept in a floor space of 38 cm x 75 cm cages; individual breeding bucks should be housed in 75 cm x 75 cm cages. Wire bottom cages which allow the feces to fall through are superior to solid bottom cages as they help control parasites (Figure 4). The wire between should be 2.5 cm x 5 cm wire mesh or 1 cm hardware cloth.The hutch dimensions can vary, but the depth should be no longer than the owner's arm length. Generally, 75 cm wide and 91-122 cm deep is the ideal size. The roof should provide at least 15 cm of overhang to shield from wind and rain.Roof construction can be a solid sheet of wood, metal, several boards or straw thatch. If wood is used, the roof should be covered with sometype of water resistant materials. Also note that multiple cages can be built, with two adjoining cages sharing a common wall (Figure 7).

Cage doors will require hinges.These can be made of metal, leather, canvas or cloth.If materials such as hinges are not available for a cage door, cages can be built with removable tops (Figure 8).Some method, such as hook-and-eyes or bent nails, should be used to secure the top in position. Scrap lumber and 2.5 cm poultry wire mesh are useful for hutch construction. Wood-framed hutches should be built with the wire mesh covering placed on the inside of the wood frame to help prevent the rabbit from chewing up the caging (Figure 9).

All-wire cages can be used in any covered type of building, such as a pole barn or open pole shed.The all-wire indoor cage is ideal for large operations, as it helps reduce disease problems. All-wire cages can be easily sanitized by using a propane hand torch plus a steel brush. If this equipment is not available, a thorough scrubbing with soap and water followed by a chlorine bleach or week lye rinse is adequate. Periodically, following cleansing, the cages should be set in the sun for 2-3 days (see Section 7).

In order to construct an all-wire cage you need 5.5 meters of 60 cm wide by 1 cm mesh hardware cloth. Roof and floors are cut out to be 60 cm x 90 cm (Figure 10). Next remove one piece 60 cm x 90 cm for a door. The remaining wire (300 cm x 60 cm) can be shaped into a rectangle (60 cm x 90 cm and 60 cm high).The top and bottom are then attached using J-clips, or other small metal or wire fasteners. Finally cut a 45 cm square hole along one of the long sides. Attach the 60 cm x 60 cm piece as a door over the 45 cm square hole. This provides 5400 square centimeters of floor space. A doe with young rabbits requires more floor space (6500-7500 square centimeters), especially if the young rabbits are left with the doe until they reach butchering size (2 to 2.5 kilograms).

Remember, any fairly durable material such as thatching or small diameter tree branches may be used to construct hutches of appropriate size. If these materials are used, they must be frequently inspected and changed as needed for chewing-damage and sanitation.

Keep outdoor cages raised off the ground and note the placement of buckets of oil around the base of each leg and/or metal cones, which can be attached around the legs (Figure 5). These tips help prevent access by rats, mice, other rodents, snakes and crawling insects. In warm areas, an open sided pole barn is best to allow adequate air circulation around the rabbit cages (Figure 11). If other livestock are housed near the rabbitry, the rabbit should be screened from the livestock. A pig or goat may frighten a nervous doe and cause her to abandon her litter.

In hot areas it is important to control the temperature. When the temperature in the hutch exceeds 30ºC it is essential to provide additional cooling measures. One simple method of shading the rabbitry is to make an awning framework covered with thatched material, moistened cloth or moistened burlap feed bags. Of course, trees and vines should be used whenever possible to shade the rabbitry. Large rabbitries in enclosed buildings often require roof sprinklers and electric fans; however, these techniques are usually expensive.

Gravel or sand floors under the caging is recommended. Loose dirt will also work to help reduce humidity and ammonia levels. Concrete under caging is not recommended unless it is to be cleaned several times daily. However, concrete can be effectively used for walkways between the rows of cages in the larger rabbitry.

For temperate climates, some people have tried a combination of an underground housing area combined with an attached outer wire cage environment. The overall facility design is shown in Figure 12. This design has several advantages in that it provides safe housing from predators and it allows for a more constant environmental temperature. By growing plants on the tops of the rabbitry and utilizing the cooling effect of the earth it is possible to maintain summertime temperatures which are several degrees cooler than the outside temperatures. This design allows a burrowing animal to choose their environment. They can lay in the sun or they can retreat into the underground housing.

To construct such a facility, you would need to follow these procedures. First, a brick substructure should be built. The dimensions for the Doe/Buck pens is 76 cm deep, 30.5 cm wide and 28 cm high. The weaning pens are a little wider as they should be 76 cm deep, 48 cm wide and 28 cm high. These dimensions can vary with the size of the brick available in your area and should not be used as exact measurements. A schematic view of the construction of these brick sub-structure pens is shown in Figure 13. Note that the planter tops are mounted over the top of the sub-structure pens and that these sub-structure pens are then attached to the outer wire cages which may be suspended on a bamboo pole in front of these pens. An alternative would be to construct a more permanent framework on which these wire cages can be mounted. This design is the one shown in Figure 12. The planter boxes are constructed with an 8 cm edge around a plastic covered wood bottom which has been sized to fit safely on the brick sub-structure. These boxes are then filled with compost from the rabbit houses and used to grow plant material for the rabbits. It is important to slope these planter boxes so that rain water will drain toward the back of the house. The inner brick wall should be at least one brick higher than the surrounding soil so that drainage from these planter tops does not flood the sub-terranean pen. Another cross sectional view of this arrangement is shown in Figure 14.

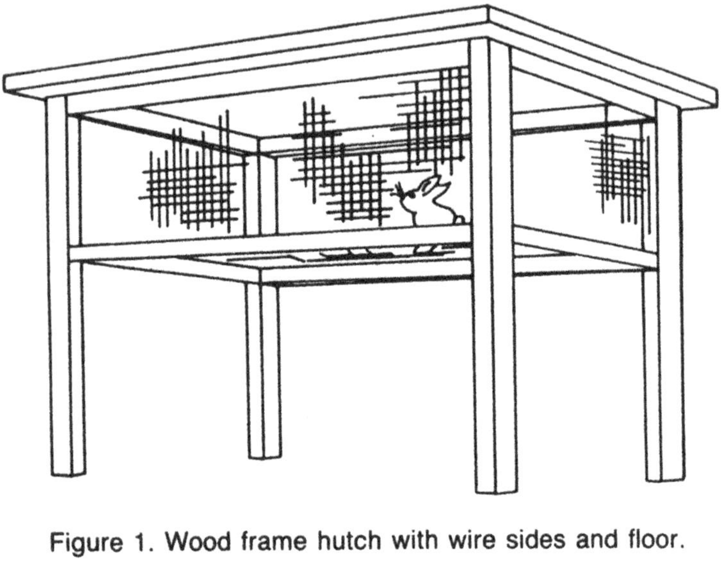

Figure 1. Wood frame hutch with wire sides and floor.

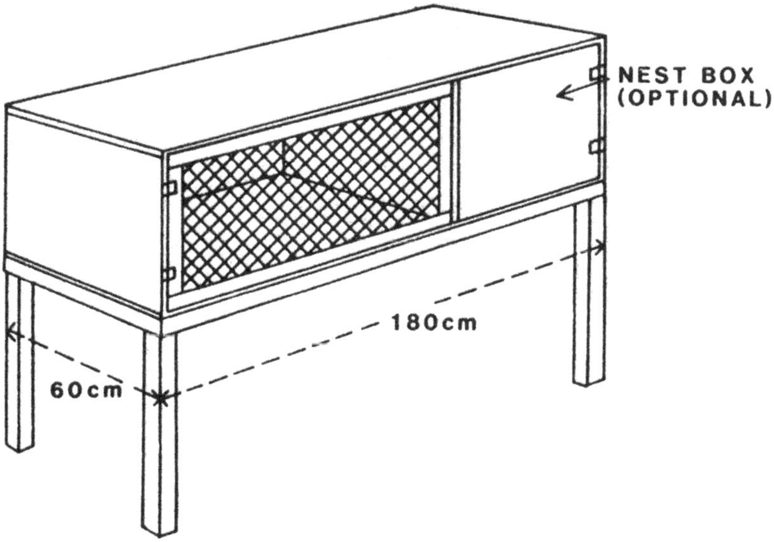

Figure 2. Wood frame hutch with solid floor and nest box.

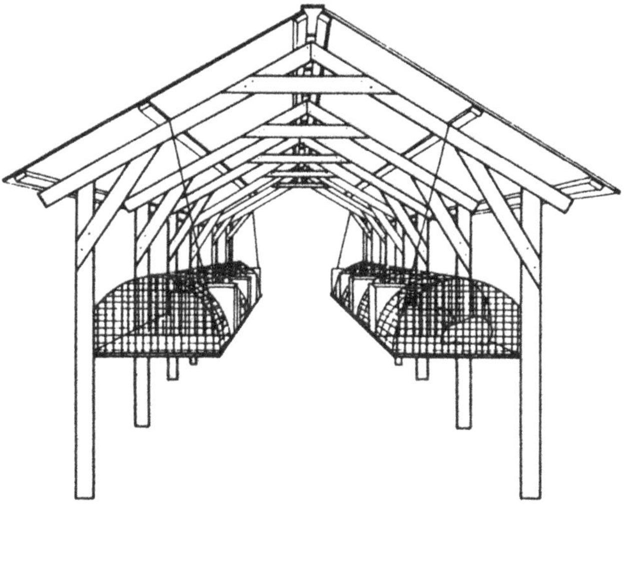

Figure 3. Hanging wire quonset style cages in open building.

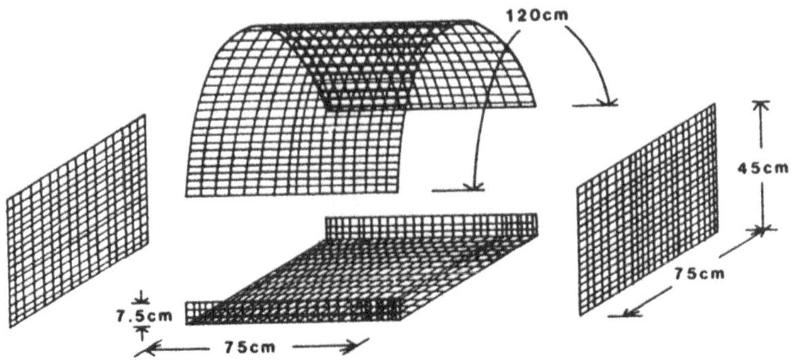

Figure 4. Construction of quonset style rabbit cages.

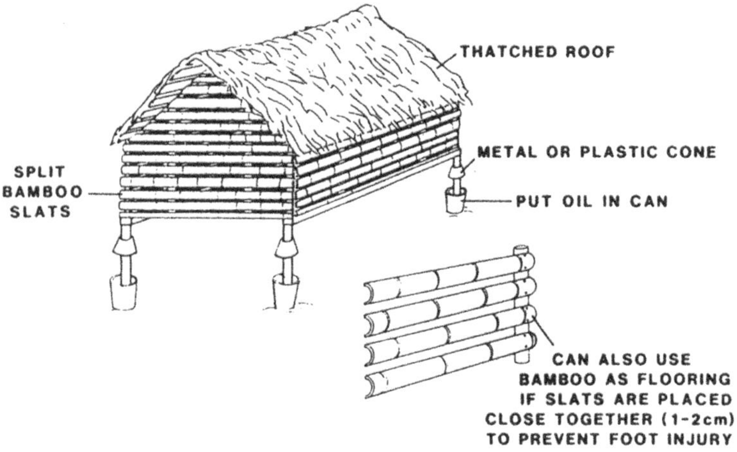

Figure 5. Bamboo hutch with a thatched roof. Note the protective barriers on the legs to keep out rodents and snakes.

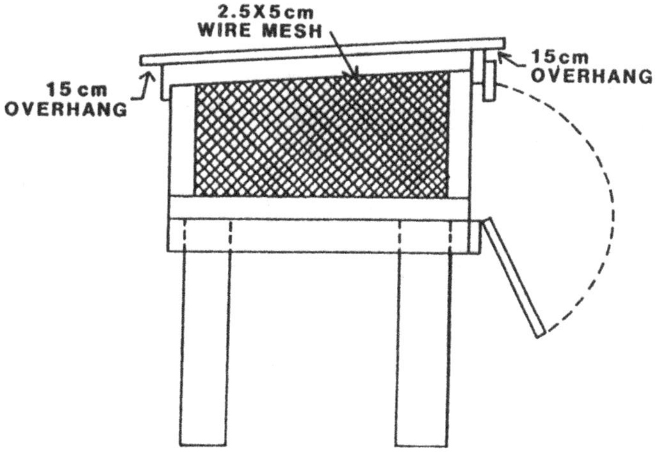

Figure 6. Wood frame hutch with wire mesh sides and roof overhang.

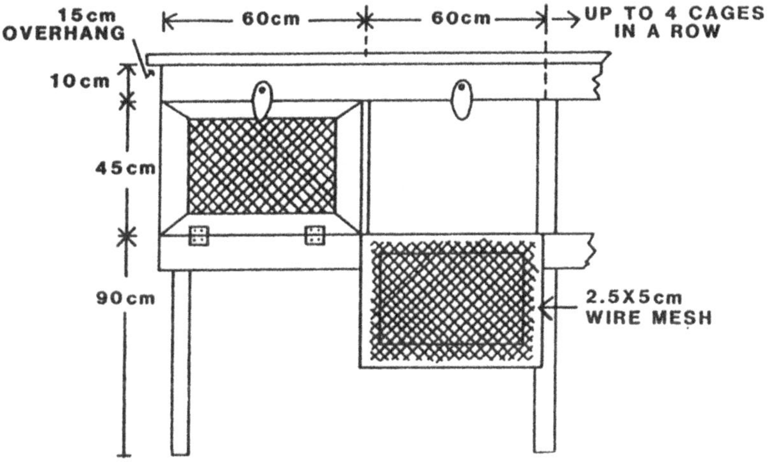

Figure 7. Model illustrates multiple hutch construction.

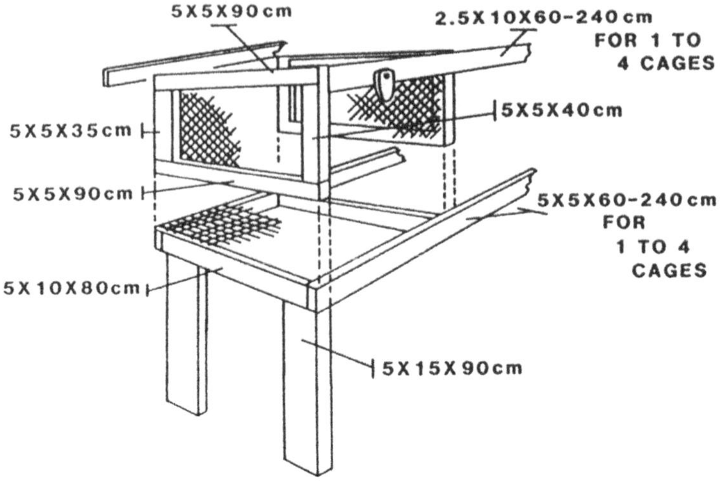

Figure 8. Model illustrates hutch construction with removable top.

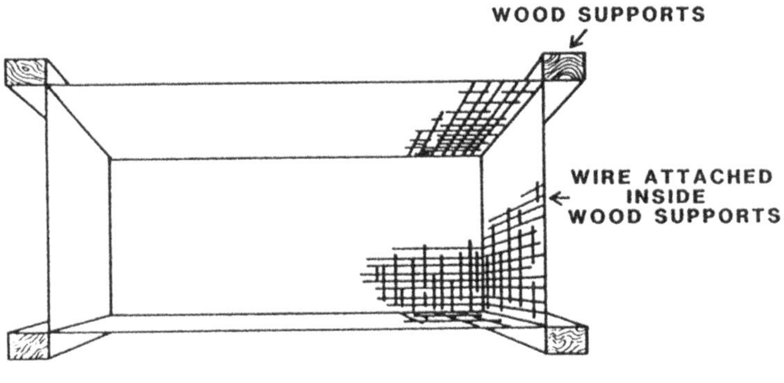

Figure 9. Overhead view of cage showing position of wire.

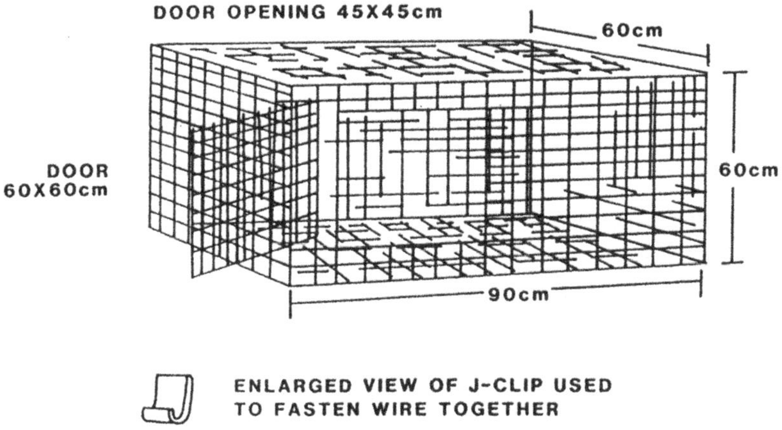

Figure 10. All wire cage construction.

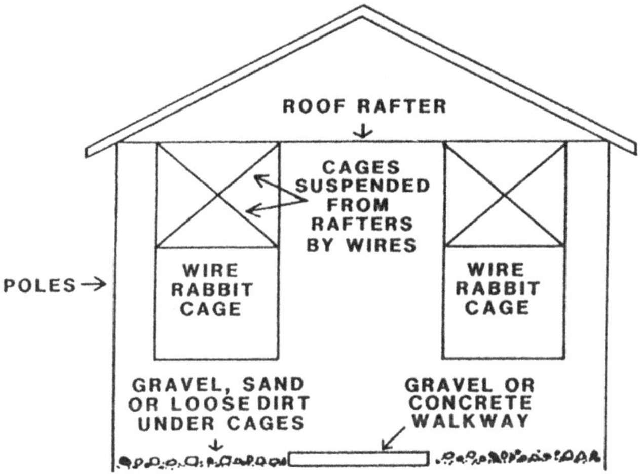

Figure 11. Open-sided pole barn with hanging cages.

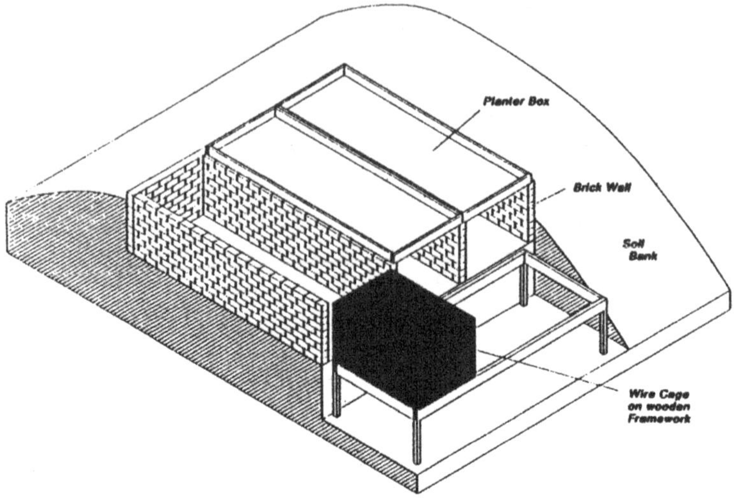

Figure 12. Underground rabbitry.

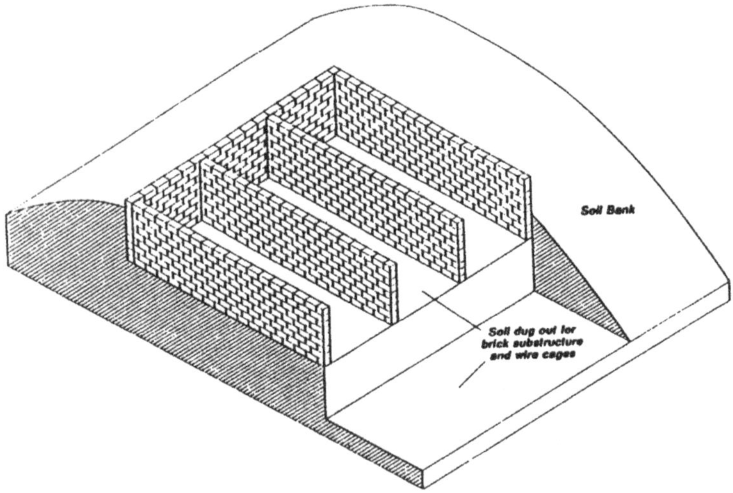

Figure 13. Subterranean brick housing.

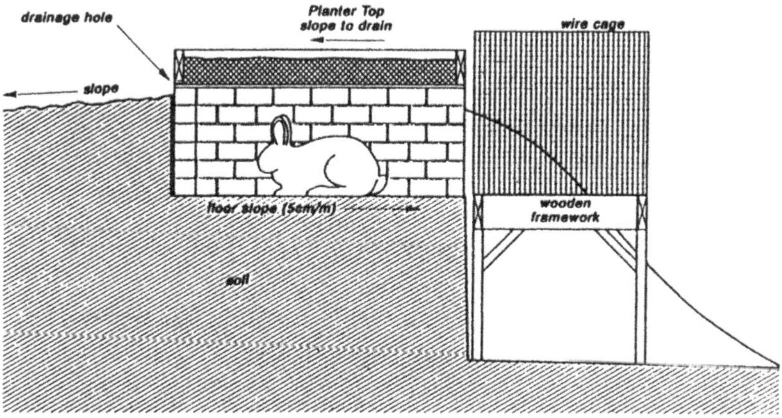

Figure 14. Cross-section view of underground rabbitry.

The underground floor of the rabbitry is covered with a mat of bamboo which can be woven together with vines, string or fine rope. It should be recognized that the more resistant the material is to chewing and rotting, the longer the cage will last. If fine mesh wire is available, this can be placed on the floor to prevent the rabbits from chewing. Whatever is used should be smooth and anything that has sharp edges should be avoided as this will only result in sores on the feet and legs. It is also good if the floor of the sub-terranean housing can be sloped so that there is drainage away from the housing unit. Building this housing unit on a piece of ground with a gentle slope to it will allow for ease in cleaning and avoiding standing water in times of heavy rain. The outer wire cages can be constructed as discussed in earlier designs. The dimensions for these cages could be 56 cm deep, 46 cm wide and 40 cm high for the Doe/Buck cages. The weaning pens could be 56 cm by 56 cm wide by 40 cm high. These are only suggestions for sizes, however, healthy rabbits do like to have room to exercise, so do not make the pens too small.

FURNITURE

Nest Box:

There are a number of nest box designs, but basically all designs try to provide the doe with some seclusion for kindling (giving birth to the young) and to provide protection for the young rabbits. The nest box should be large enough to prevent crowding, but not too large to prevent the young from remaining close together. Good ventilation is required in warmer climates, while some protection from wind is required in cooler climates.

Box-type: The box-type nest box can be constructed of a variety of scrap wood material such as shipping crates, etc. The corners should be covered with metal if possible to prevent the doe from gnawing on these areas. Basic dimensions shown in Figure 15 are for a 3.5-5 kg doe. Adjustments in the size can be made for larger or smaller does. A good doe should provide a warm nest of wool groomed from her body or from nesting material provided (such as straw), so that her young can survive temperatures as low as -25°C. Lids and floors should be secured in place, but in a manner that allows the nest box to be removed and cleaned between litters (Figure 16). Extra insulation can be provided by placing corrugated cardboard or paper as a liner along the inside of the nest box. An alternate method is to place the nest box within a corrugated cardboard box and use straw to line the space between the nest box and the cardboard box.

Nail-keg: The nail-keg (or any small wooden barrel) nest box can make use of an item sometimes available for free in the community. A 30 cm diameter keg is required for a doe weighing over 5.5 kg, while a 25 cm diameter keg is adequate for a 3.5 kg doe. The required drain holes, ventilation holes and door are shown in Figures 17 and 18. Boards nailed to the front and rear of the keg will prevent it from rolling.

Open-box type: When plywood is available, a more open wood nest can be built for use in warmer climates (Figure 19). The bottom can be either solid or wire window screen material. The screen bottom has the advantage of keeping both the babies and the doe cooler by providing more ventilation. If a solid bottom is used, 3-4 holes should be drilled in the bottom to provide drainage. The sitting board at the back of the box provides the doe with a place to rest and observe her young without having to be in among them.

Wire nest box: In some tropical areas the sides and bottom of the nest box can be constructed entirely of wire and the top is left open. The purpose of this type of nest box is to provide adequate ventilation while confining the young to one small part of the cage.

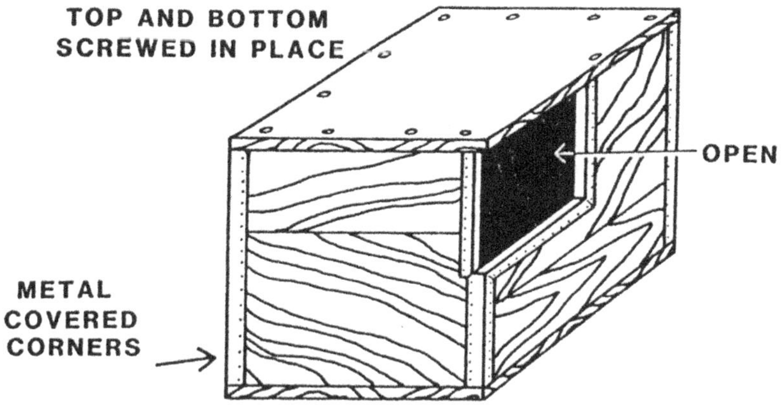

Figure 15: Nest box construction using wood with metal covered corners.

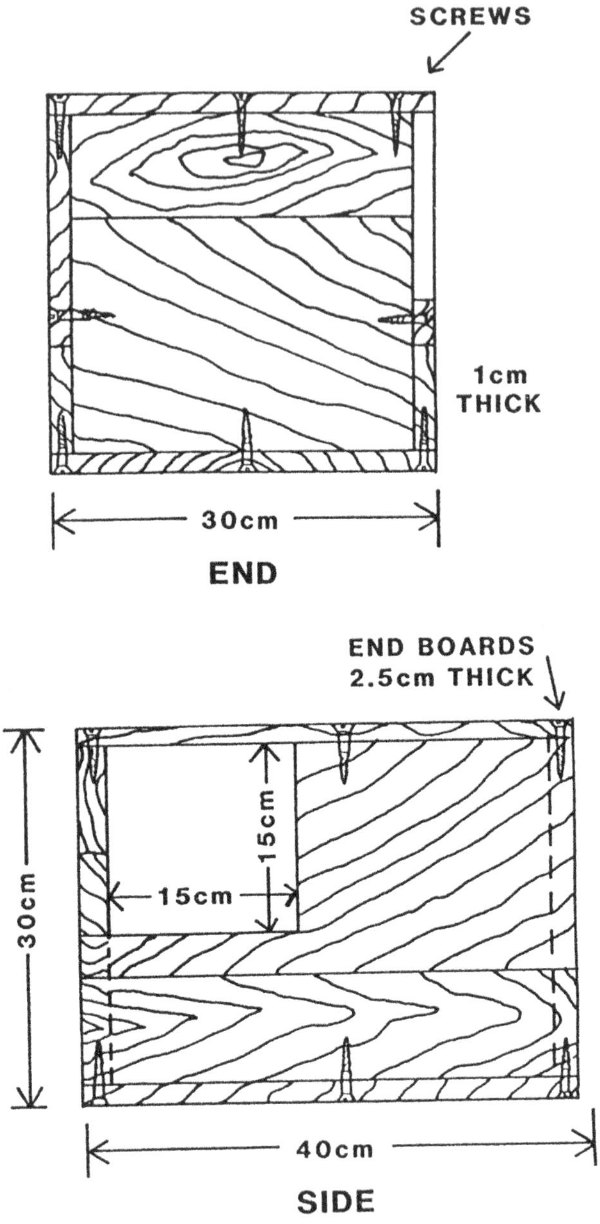

Figure 16: Nest box construction detailed for ease in cleaning.

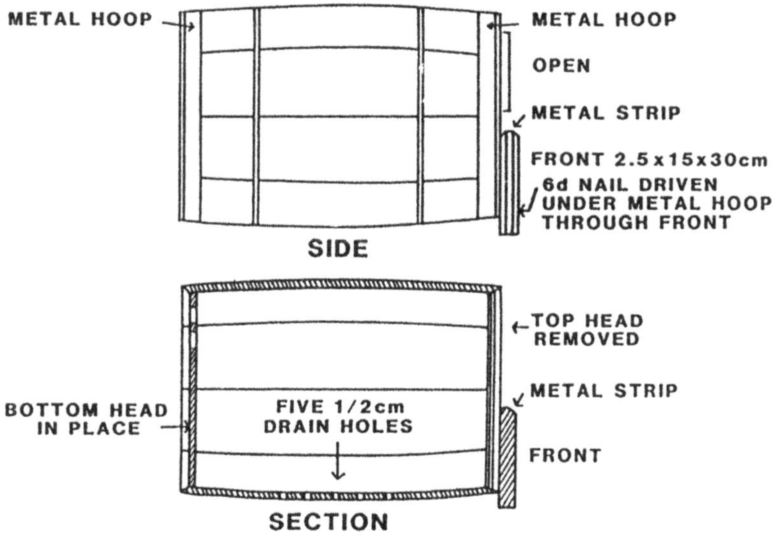

Figure 17: Nail-keg hutch.

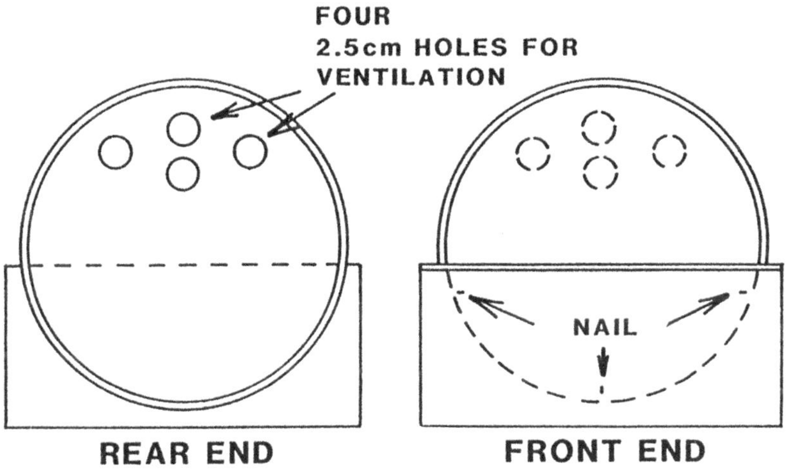

Figure 18: Nail-keg hutch with wood supports to prevent it from rolling.

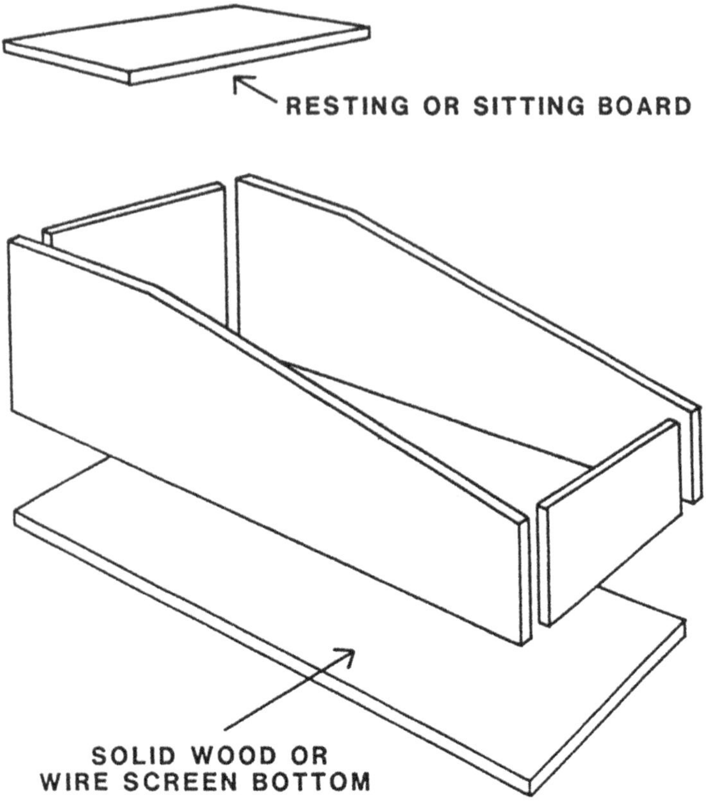

Figure 19: Open wood nest box.

WATERING DEVICES

Rabbits require clean water at all times. A doe and her offspring can drink four liters of water daily. Several watering systems, from simple equipment for small farmers to automatic systems for larger operations, can be used.

Crocks:

Two liter pottery crocks (purchased or hand made) are still one of the easiest and least expensive ways for watering rabbits. The crock needs to be fastened to the front of the hutch with part of the crock extending into the hutch. Fasten the crock so that the rabbits will not tip it over. Clean and disinfect the crocks at least weekly (see Section 7).

Coffee Cans:

Cut-down coffee cans or other types of large cans will work in place of crocks, but are tipped over more easily by the rabbits. Therefore, it is necessary to securely fasten them to a board or the cage. Again, if part of the can extends out beyond the cage it can be refilled easily. Cans should be cleaned and disinfected weekly, and replaced as soon as they begin to show rust.

Soda Bottle and Cans:

A further modification of the can watering device is to wire the can inside the hutch. Then wire a large soda bottle on the outside of the hutch with the bottle neck extending into the water in the can (Figure 20).

Plastic Jugs:

A simple watering device can be made from a four liter plastic jug such as those used for various cleaning agents (bleach or soap). Be sure to rinse the container thoroughly, then mount a drinking valve in the heavy plastic material near the bottom of the jug (Figure 21). The jug is then placed on a tray attached to the cage. A wire loop helps secure the jug in place. By retaining the jug's plastic screw-cap, the water supply can be protected from contamination by dust or vermin. This system is also useful for medicating a single rabbit. The medication is added to the water in a measured quantity and there is little waste as encountered in the more open systems described previously. The drinking valve would have to be purchased from a rabbit or poultry supply company.

Automatic-Watering Systems:

Automatic systems eliminate the need for washing and refilling of water containers daily and deliver a constant supply of clean, fresh water. The system starts with a pressure-reduction tank on the incoming water line. The tank is equipped with a float valve feeding into a supply pipe that leads to the watering unit for each hutch. The tank should be about 30 cm higher than the highest point in the piping system in order to maintain a good water flow. Tanks as small as four liters can be used where there is no problem with sediment in the water and no danger of freezing in cold weather. Small tanks have the advantage of maintaining a cool, fresh water supply due to the constant turnover. The tank should be connected to the watering pipes by a flexible hose in order to allow for adjustment in tank height. If the tank is too high, the system will be under too much pressure and the rabbit will not be able to trip the valves in their hutches; and, if the pressure is too low, the valves will leak or drip.

The watering valve for the rabbit should be located 22.5 cm above the hutch floor for large breeds and 18 cm above the floor for smaller breeds. To minimize problems from dripping water valves, locate the valves just outside the hutch and cut a hold in the hutch wall large enough to allow the rabbit to stick its head through and drink.

The piping from the tank to the cages can be either a flexible rubber-type hose such as a garden hose or plastic (PVC) piping (except the first section near the tank as noted above). Drinking valves would have to be obtained from a commercial source of rabbit or poultry supplies. Additional valves for bleeding air from the system should be located at any high point in the piping. A valve to drain the system should be located at the system's lowest point (Figure 22).

Rabbits quickly learn to use the system. Even young rabbits readily adapt to the system providing they are large enough to reach the drinking valve. If young rabbits are not large enough to reach the drinking valve, be sure to supply their fresh water needs by adding a crock or other device they can reach. When a hutch has been unoccupied for a few days, make sure the drinking valve is working properly before introducing a new rabbit, as minerals and sediment in the water can cause the valve to stick.

In cold climates you must protect the system from freezing. If the hutch and all piping are in a heated enclosure there should be no freezing problem; otherwise, heating cables sold for wrapping house pipes can be used on the rabbit piping. Another alternative for short cold periods is to keep the water in the system moving by allowing water to dribble from the drain valve at the low end of the system.

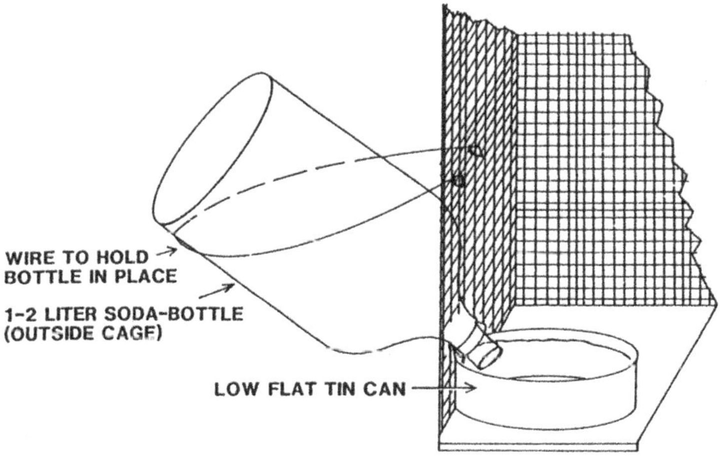

Figure 20. Cut-a-way view of soda-bottle waterer.

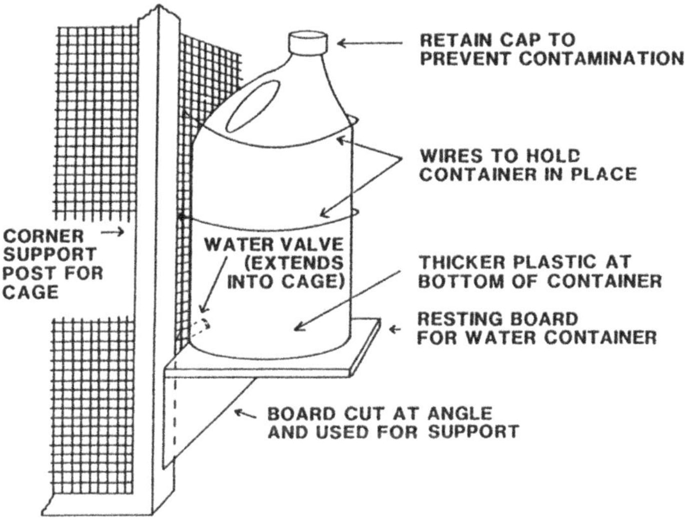

Figure 21. Plastic jug watering device.

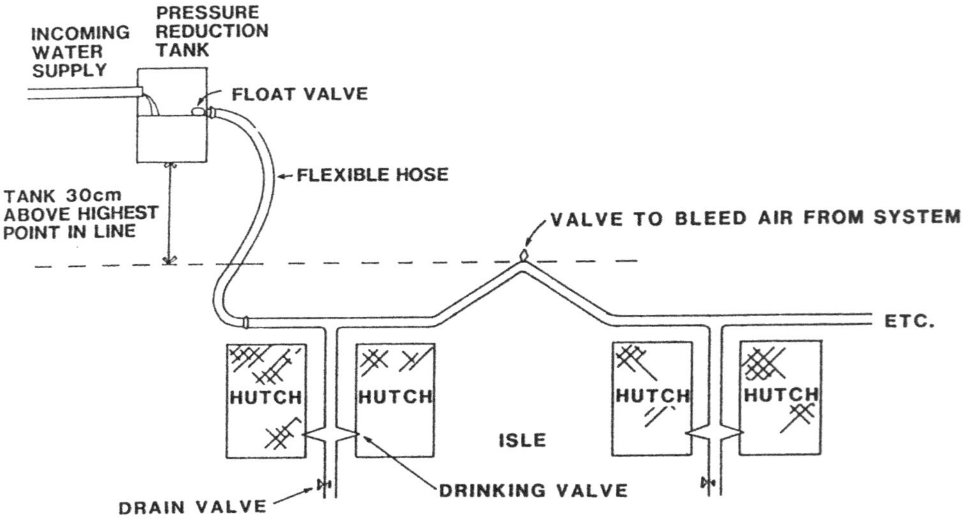

Figure 22. Automatic watering system for large rabbitry.

FEEDERS

Feeders vary from small, homemade devices to manufactured hoppers for pelleted food. For the small rabbit farmer, homemade feeders are the most practical and least expensive alternative. The simplest feeders are those using a pottery crock or tin can similar to the equipment used to hold water.

Tin can feeders:

Take a small coffee can or smaller size tin can, and with tin snips cut down one side to within 5 cm of the bottom. Next, cut half-way around the can staying parallel with the bottom (Figure 23).Now flatten the cut section and fasten the overlap inside the remaining section of the can using rivets or solder. Bend the lower portion of the can over to reduce the possibility of the rabbit cutting itself on the can.The feeder can be wired to a cage wall to prevent tipping (either inside or outside the cage, Figure 24).

Feed troughs:

Hay mangers and feed troughs keep food off the floor of the cage where it can be trampled and lost. Mangers and troughs also save time in feeding. A simple feed trough (Figure 25) can be built from wood and metal. If available, 24 gauge galvanized metal should be used as this material is more resistant to rust and corrosion. The trough guide is made from two wood blocks approximately 2 cm thick (9 cm x 21 cm in length and width) and covered on the flat surface with pieces of metal (to stop chewing). Next, cut two pieces of metal 11.75 cm wide and 30 cm long or longer. The length of these pieces can vary and will determine the length of the feeder, but both pieces should be the same length.Now cut five pieces of 2 cm x 4 cm boards to 30 cm length (or the same length as the previous pieces). The 2 cm x 4 cm boards are attached at their end points to the metal covered 9 cm x 21 cm board using wood screws (Figure 25). Then the 11.75 cm x 30 cm metal pieces are formed into two half-circles and placed between the 2 cm x 4 cm x 30 cm pieces (Figure 25). These two half-circles can be attached to the wood 2 cm x 4 cm supports with wood screws. A guard to prevent small rabbits from sitting in the feed trough can be constructed from a piece of iron 0.5 cm x 2 cm x 30 cm to which 0.5 cm iron rods are attached at 7.5 cm intervals. The wire rods need to be 17 cm long to cover the feed trough area, and are attached at their midpoint to the flat iron pieces (Figure 26). The flat iron piece is then attached with wood screws to the center 2 cm x 4 cm wood support for the two troughs.

Hay manger:

A manger can be constructed from 26-gauge (galvanized, if available) metal and 16-gauge 2.5 cm mesh wire (Figure 27).Metal sides and ends are attached to the 16-gauge 2.5 cm mesh wire with 2.5 cm long metal clips and rivets (pop rivets or similar devices).Hay can then be placed in the manger and not be contaminated by feces and urine, as it would be if just placed in the bottom of the cage. If the length and width dimensions used for the manger are the same as those used for the feeding trough, the manger can sit on top of the trough thus conserving cage space (Figure 28).

Oil-can feeder:

A four compartment feeder can be constructed out of a five gallon (20 liter) oil or gas can.The can should be cleaned thoroughly, then two openings are cut in the front and two in the rear(Figure 29).The openings are 7.5 cm x 10 cm and are located 10 cm from the bottom of the cage. Hem (turnover) a 0.5 cm section around the edge of each hole to create a smooth surface. The inner partitions can be built of wood. Two partitions the same size as the inside dimensions of the can are needed (Figure 30). Notching these in their center for half their length will allow them to interlock.The four angled pieces can be made now, also taking their dimensions from the inside compartments. The cut-out portions of these pieces allow food to drop down to the feeding area inside the openings. A piece of 26-gauge (galvanized, if available) metal can be used to build a top for the feeder.The corners are cut and turned down for 2-1/2 cm (Figure 31). If an oil or gas can is not available, the feeder can be constructed entirely of galvanized metal.

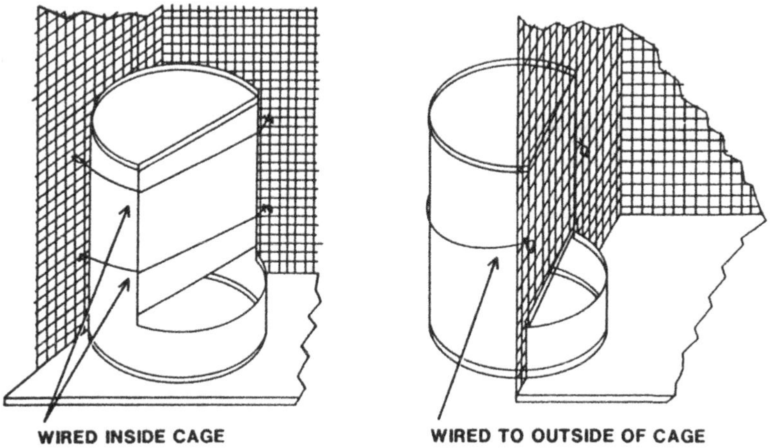

Figure 24: Tin-can feeder attached to cage or hutch.

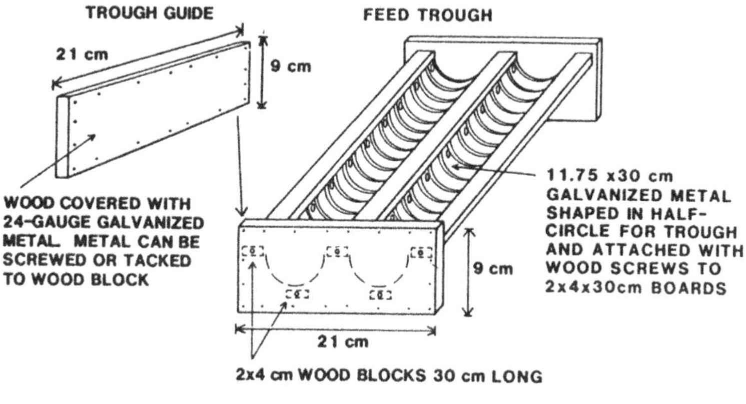

Figure 25. Wood and galvanized metal feed trough

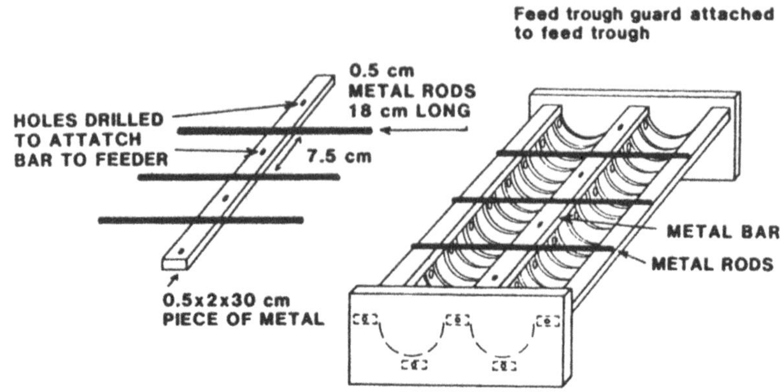

Figure 26. Feed trough guard.

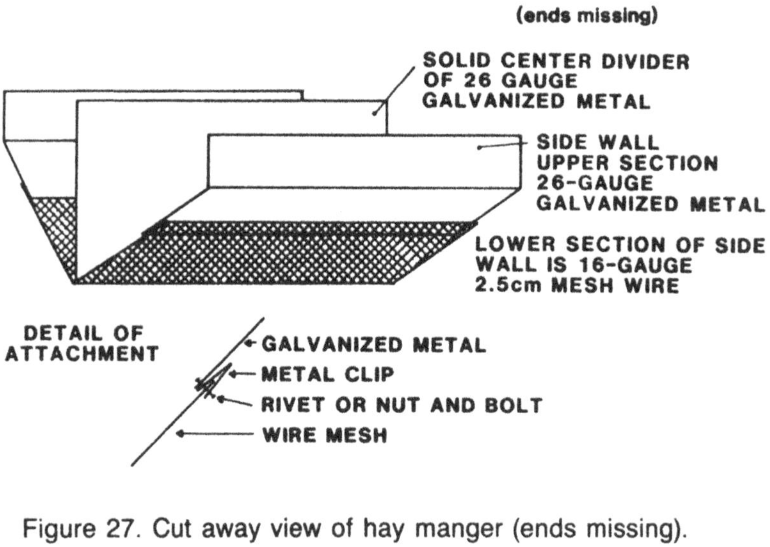

Figure 27. Cut away view of hay manger (ends missing).

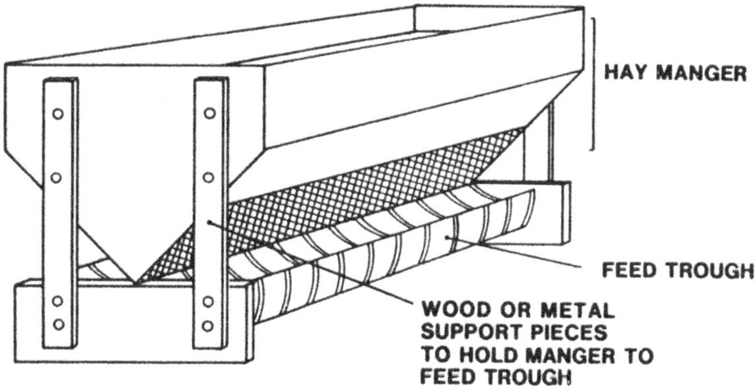

Figure 28. Hay manger on top of feed trough.

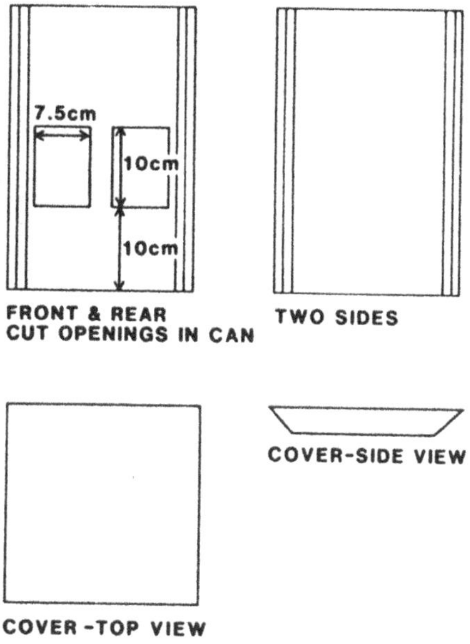

Figure 29. Self-feeder made from five-gallon oil can.

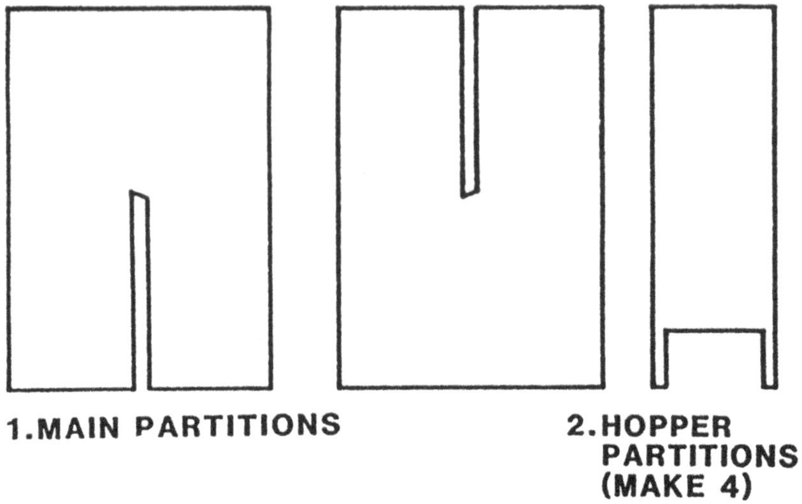

Figure 30. Partitions for five-gallon self-feeder.

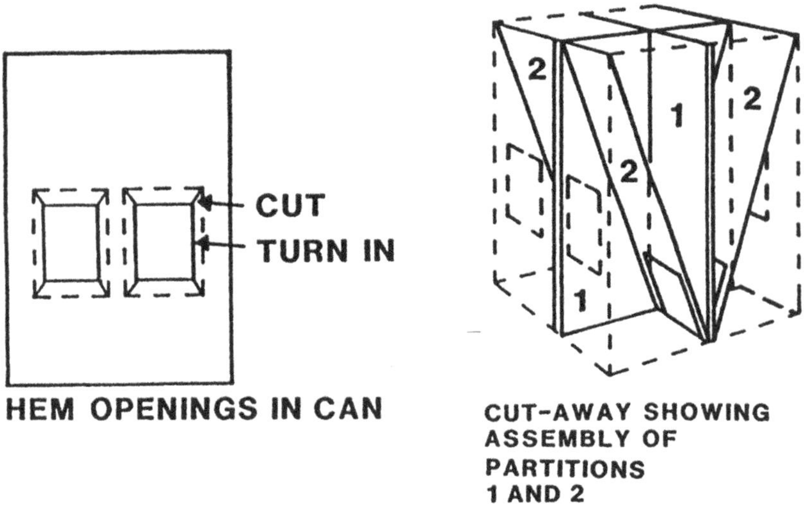

Figure 31. Oil can feeder partitions and openings.

CARE OF RABBITS DURING EXTREME HEAT

Recall that the domestic rabbit (Oryctolagus cuniculi) originated in Europe, and thus is best suited for cooler temperatures. If the ambient temperature in your area is greater than 25-30° C, then heat stress may be a problem to consider when constructing housing. During warm weather, the two most important requirements for rabbits are shade and good air circulation. In addition, rabbits should have a good supply of fresh water. As water intake often dramatically increases during warm weather, providing fresh water ad libitum (free-choice) is preferred.

There are several methods of decreasing the temperature of the hutch housing a rabbit.Shade is provided by constructing buildings with solid tops. Aluminum, or another reflective (shiny) metal helps reflect the heat of the sun. Open-sided buildings allow wind to move through a rabbit barn. Fans, either electrical or manual, also can provide circulation of the air around rabbits. Air movement increases the release of internal heat from a rabbit, thus decreasing the body temperature.

Evaporative cooling, such as is provided for humans by perspiring (sweating) may also be used to decrease the temperature of the air around an entire hutch. Overhead sprinklers, which spray a fine mist of water over the hutches, can be constructed with traditional plumbing methods or simulated by gently scattering water over the tops of the hutches. When using an overhead sprinkler system, wet the tops and any solid sides of the hutch.The resulting evaporative cooling will reduce the temperature in the hutch by 3-8°C. Even a system as simple as soaking burlap, canvas, or other cloth in water and hanging from the edges of the roof will shade and cool the air.

Newborn liters and pregnant does, especially those near kindling, are most susceptible to heat stress. Signs of heat stress range from restlessness in early stages, to rapid respiration, excessive moisture around the mouth, and small hemmorages (sites of bleeding) on the nostrils. Place a rabbit suffering from heat stress in a quiet, cool, well-ventilated location. Wet the rabbit or give it a cloth soaked in water to lie on. See Section 6, "Diseases", for more information.

Section 3

Handling and Restraint And Normal Physiological Values

HANDLING, RESTRAINT AND NORMAL PHYSIOLOGICAL VALUES

HANDLING AND RESTRAINT

Unlike other small mammals, the rabbit has a significant proportion of its weight distributed in the rear portions of its body. Heavy musculature for running is found in the lower back and hind limbs. Rabbits also have a light and somewhat delicate skeleton. They are therefore quite susceptible to fractures of the vertebrae (broken backs). See Section 6, "Miscellaneous and Noninfectious Diseases". Proper handling and restraint is therefore very important.

The best way to carry a rabbit long distances is to place them in a small dark box or bag, although they can be carried short distances by hand. Pick up a rabbit by the "scruff" of the neck, similar to the way an adult female mammal picks up her offspring. It is very, very important to support the rear limbs with your free hand. One quick, powerful kick by a "scruffed" rabbit with unsupported rear limbs is enough to fracture its back. Hiding the face of the rabbit with a small cloth or in the crook of your elbow often quiets a rabbit so that they do not struggle while being moved. See Figures 32 and 33 for proper restraint and handling.

Rabbits' ears appear to be quite convenient "handles". However, these structures are very delicate, sensitive and should **never** be used to pick up a rabbit.

When restraining a rabbit on a table for examination, it is best to place the rabbit in a natural position and encircle it with your arms. As mentioned earlier, covering the eyes of a rabbit often quiets it down significantly. One method that has been used to "hypnotize" rabbits consists of turning the rabbit on its back. The head is held in a flexed position with the chin tucked in tight against the neck. The hind limbs are then grasped with the other hand and the body is stretched out very tightly. After holding the rabbit in this position for 1-2 minutes, release the hind limbs and gently stroke the abdomen from the chest to the end of the extended hind limbs. This has been used for minor procedures such as examination and injections. Loud noises will awaken the rabbit from this state of somnolence. Also, release of the head from its flexed position will result in the immediate awakening of the rabbit.

Figure 32. Rabbit restraint - lift a rabbit with two hands and always support the posterior end of the body.

Figure 33. Proper method for carrying and supporting a rabbit to prevent injury.

NORMATIVE PHYSIOLOGICAL VALUES

The following table (Table 1) provides an easy reference of information on several aspects of rabbit biology and physiology. Note that these values are based on the domestic rabbit breed New Zealand White of the species Oryctolagus cuniculi and may be different from other breeds and species, although close similarities most likely exist.

Table 1
NORMATIVE PHYSIOLOGICAL VALUES

MEASUREMENT/ COMPONENT	RABBIT New Zealand Whites *Oryctolagus cuniculus*	MEAN (average)	MEAN ——— Male	MEAN ——— Female
Recommended room humidity	50%			
Recommended room temperature	(68°F) 20°C			
Begin consumption of dry feed	21 days			
Daily food consumption (adult)		145 g (about 4.5% of BW)		
Water requirements	200-850 mL/day			
Adult male, body weight (BW)	2000-7000 g	3000 g		
Adult female, body weight (BW)	2500-8000 g	3500 g		
Female breeding age	5-6 months			
Male breeding age	6-7 months			
Estrous cycle	Polyestrous			
Gestation	28-35 days	31 days		
Birth weight	80-120 g			
Litter size	1-18	7-9		
Weaning age	8 weeks			
Life expectancy (years)	5-12 years			
Respiratory rate (breaths/min)	32.0-60.0	51.0		
Tidal volume (mL)	19.3-24.6	21.0		
Minute volume (L/min)	0.37-1.14	1.07		
Whole blood volume (mL/kg BW)	55.6-57.3			
Plasma volume	(mL/kg BW)	27.8-51.4	38.8	

Table 1
NORMATIVE PHYSIOLOGICAL VALUES

MEASUREMENT/ COMPONENT	RABBIT New Zealand Whites *Oryctolagus cuniculus*	MEAN (average)	MEAN — Male	MEAN — Female
Erythrocyte volume (mL/kg BW)	16.8-17.5			
Whole blood pH	7.21-7.57	7.35		
Heart rate (beats/min)	306-333			
Body temperature	38.6-40.1ºC	39.5ºC		
Erythrocytes (RBC) ($x 10^6$/mL)	4-6.7		6.7	6.31
Hemoglobin (Hb) (g/dL)			13.9	12.8
Mean corpuscular volume (MCV) (fL)			62.5	63.1
Mean corpuscular hemoglobin (MCH) (pg)			20.7	20.3
Mean corpuscular hemoglobin concentration (MCHC) (g/dL)			33.5	32.3
Hematocrit (PVC) (%)	39-42		41.5	39.8
Sedimentation rate	2.0	1.75		
Platelets ($x 10^3$uL)			480	450
Leukocytes (WBC) ($x10^3$/mL)	5.2-12		9.0	7.9
Neutrophils (%)			46.0	43.4
Eosinophils (%)			2.0	2.0
Basophils (%)			5.0	4.3
Lymphocytes (%)			39.0	41.8
Monocytes (%)			8.0	9.0

Section 4

Reproduction and Genetic Considerations

REPRODUCTION AND GENETIC CONSIDERATIONS

INTRODUCTION

In planning a breeding program for rabbits, concepts that have been shown to work in the breeding of other domestic livestock may be applied in the rabbitry.

There are very few studies on the genetic parameters of reproductive traits in rabbits. However, litter size at birth is probably the most important trait in the reproductive performance of animals which give birth to more than one offspring, such as rabbits. Some people feel that litter size and weight at 21 days of age are very good indicators of milk production and maternal ability. Young rabbits do not eat solid food or drink water before this age and, consequently, their survival and weight are due only to the mother's milk and maternal ability. Litters are usually weaned by 28 days of age. Thus, knowledge of the heritability of these traits is very important in a genetic improvement program.

Gene Pool:

The initial foundation stock provides a pool or group of hereditary units commonly called genes. Genes are found on specific locations on the chromosomes (small thread-like bodies found in the center or nucleus of every cell). There are 22 pairs of chromosomes in the rabbit, one chromosome of each pair comes from each parent (via egg or sperm). Hereditary characteristics are transmitted from one generation to the next by this method.

Selection:

Selection is the process of a planned breeding to produce offspring with a desired trait or traits. Selection depends upon the initial quality of the gene pool and a proper environment to allow the animal to grow to its maximum potential (i.e. an environment that has adequate food and water plus adequate cage space, so that the young rabbits are not stunted).

Although several breeds have been used for meat production in many areas of the world, the New Zealand White and the Californian are perhaps the two most commonly selected breeds used in subtropical areas. Studies comparing does from these two breeds have shown that breed does not appear to be an important selection factor. It would also appear that sex differences are not important for growth and carcass traits. While the Californian breed does have been shown to have a slight weight advantage at eight weeks of age, the most significant weight gains have come from crosses with the Flemish Giant breed. Studies have also shown that a doe's productivity in number of offspring will usually increase from the first to the second litter. At that time the litter size usually remains stable until the seventh litter after which productivity declines. Most breeders will cull (remove from the breeding group) the doe if she produces two litters with five or fewer kits. Does are also culled if they fail to conceive during the 7 to 17 days postpartum breeding period.

Outbreeding or Crossbreeding:

Outbreeding or crossbreeding is the mating of unrelated rabbits.

Advantages: Offspring are usually more fertile and they display increased vigor and growth (desirable in animals for meat production). It produces a maximum number of unlike genes in one animal.

Disadvantages: Later generations show a great amount of variation and can have little commercial value.

Inbreeding:

Inbreeding is the mating of closely related individuals.

Advantages: It maintains uniformity of the rabbit population over several generations.

Disadvantages: It can result in undesirable recessive genes causing a decrease in viability and reproductive fitness of future generations of rabbits. Sterility, mortality and undesirable abnormal variation occur in greater numbers.

Generally, it is desirable that the initial stock of rabbits be of the highest quality and contain the maximum number of favorable genes. If the size of the rabbitry is small, and especially if there is only one buck used, it would be highly desirable to bring new, unrelated females into the rabbitry to periodically increase the gene pool and avoid extensive inbreeding.

PHYSIOLOGY OF RABBIT REPRODUCTION

Heat (estrus):

Rabbits do not have a regular estrous cycle as do some domestic animals. Rather the doe (female rabbit) remains in heat (estrus) for long periods unless she is bred. The follicles on the ovary remain active for 12 to 16 days before regressing. When these active follicles regress, others become active to take their place. There may be a short period (1 to 2 days) between regression of one set of follicles and the onset of activity of another set when the doe is sterile.

Breeding of rabbits is a relatively easy task as the female is usually receptive when placed with the male. However, a sexually receptive doe will display some characteristic signs of estrus. These signs will include a congested vulva (appears red and swollen), she is restless, tries to join other rabbits in adjacent cages and she will usually rub her chin. When a receptive doe is placed with a male, she will raise her hindquarters. The buck mounts the doe and grasps her body with his forelegs. Intromission is accomplished after 8-12 rapid copulatory movements and ejaculation usually occurs following the first intromission. Following ejaculation, the buck will fall off to one side and he may make a crying noise. A vigorous buck may attempt to mount again. The fluid portion of the ejaculate ranges between 0.5 and 1.5 mL and the sperm density ranges between 0.5×10^6 and 3.5×10^6/mL. If the environmental temperature is extremely high, the sperm count of the males will usually fall. In this case, it has been shown that the addition of thyroxine to the diet will help improve the sperm count, providing the elevated temperatures are intermittent.

Female rabbits are fertile for all but about 48 hours per month. The doe will drop her eggs 10-13 hours after penetration by the buck. This will allow you to breed the doe in the morning and the evening of the same day which will usually assure you of a larger litter size. Some does fail to ovulate following breeding and this may be associated with a deficiency in luteinizing hormone (LH). It has also been shown that the weight of the does must be regulated as a fat doe will be much less likely to breed.

Ovulation:

Ovulation (the process of a large ovarian follicle breaking away to become an egg) occurs about ten hours after mating. At this point, the male's sperm have moved through the female's reproductive tract to a point where they will meet with and enter (fertilize) the egg.

Development of Fetus:

The fertilized egg grows and becomes the young fetus. In highly fertile strains an average of ten eggs are shed, while in the strains of lower fertility only 4-5 eggs may be shed. The average gestation period for the rabbit is 30 to 33 days (range of 28 to 35 days).

BREEDING METHODS

Age to Breed:

Smaller breeds develop more rapidly and are sexually mature earlier. Normally bucks are one month older than does before the first breeding. In general, medium weight (4-6 kilograms) rabbits are first bred at 5-6 months and giant breeds (10 kilograms) at 8-10 months. Because the buck is very territorial, a doe to be bred should always be placed in the buck's cage. In order to prevent back injury, it is important to carry the rabbit in the correct manner (Figure 32 and 33). The rabbits should be observed for 15 to 20 minutes, and if no breeding takes place, it is best to try another buck or try again the next day. If non-receptive, the rabbits may fight.

Pregnancy Detection:

Pregnancy can be accurately determined by palpating the abdomen at 12-14 days after mating. Please follow these steps:

1. Place the rabbit on a table or other flat surface. Hold the ears, fold of neck and shoulder skin in the left hand (or right hand for left-handed persons).
2. Place the right hand under the abdomen and in front of the pelvis; place the thumb on one side and the fingers on the other side of the abdomen.
3. Slip the tissues of the abdomen between thumb and fingers, feeling (gently) for several marble-size (2 cm diameter) forms.
4. **CAUTION:** Do not palpate with too much pressure as the uterus can be damaged or a fetus torn loose from the uterine wall. This can result in a toxic abortion with loss of the fetuses and possibly the doe.
5. Experienced rabbit owners may be able to palpate pregnant does as early as seven days post mating. If by day sixteen no fetuses are palpated, the doe should be rebred.

Because they have a postpartum heat period, they can also be rebred immediately following parturition. In effect this would mean that a doe could be rebred on a monthly basis. However, since the young are usually weaned at 28-30 days of age it is best to stagger the breeding periods so that a doe will have about eight litters per year. A doe can usually be palpated for pregnancy as early as 10-12 days. By gently compressing the walls of the abdomen toward the midline between your fingers it is possible to feel small enlargements. These are usually foeti, however, these must be distinguished from fecal matter in the intestine, enlarged mammary glands and retained placental material.

Kindling (giving birth to young):

Parturition usually occurs in early morning hours and this is called kindling. Several days prior to kindling, the doe will collect nesting material (hay, straw, excelsior). The doe will then begin to pluck her hair and line the nest. As the young are born, the doe eats the placenta and fetal membranes. She will also sever the umbilical cord to each fetus. Some abnormalities to look for include failure of the doe to build a nest, birth of the young outside the nest, scattering of the young, and cannibalism. Sometimes a doe will look as if she is pregnant and she will build a nest, but there are no offspring produced. This is usually a sign of a false pregnancy. In many cases the doe will return to normal cycling and breeding following one of these false pregnancies. The average litter size is eight kits. These kits should reach 10 to 12 kilograms at three months of age and this is the age at which most are slaughtered for food.

Complications at Kindling: Complications are rare. Both anterior and posterior (breech presentations) are considered normal; however, if it is determined that the doe is not having strong contractions it may be necessary to give 0.2-3 U/kg of oxytocin (Table 3). Because of the stress of kindling, does are more susceptible to diseases at this time. Pneumonia is the most frequently seen secondary problem in kindling does. Depending on the cause of the pneumonia, antibiotics may be effective in treating the condition (Table 3).

Caked Breasts: This condition develops in does if their milk is not removed. The tissues around the teats become enlarged and hard. Later the ends of the teats become tender. Treatment is to rub lanolin on the area and massage the teats, strip the milk from these teats, or allow the young to nurse on the affected teats.

Mastitis: The breast is congested and warm to the touch. Often the skin over the swollen breast is red or purple, and the teats are discolored. Usually mastitis is due to a bacterial infection and can be contagious causing illness in the nursing bunnies. Broad spectrum antibiotics can be used for treatment. Do not let young nurse on infected breasts.

Large Litters: Does may care for 9-10 young, but litters of up to 18 are possible. If there is another doe with a small litter, it is possible to take some of the young from the large litter and introduce the young to a doe with a small litter. Usually the litters should be within 3-4 days-of-age of each other.

Dry Does: The young will starve in 2-3 days if the doe has no milk to nurse them. Using a foster mother may prove useful.

Cannibalism of Young: The doe normally severs the umbilicus and ingests the placenta. Cannibalism may be due to an inadequate diet, hereditary factors, or from disturbing a nervous doe.

Weaning:

Young rabbits can be weaned at eight weeks of age, and the weight of the young should average 2 kilograms/rabbit. Weaning as early as five weeks is possible.

Production Records:

Production records can be very helpful in determining which does and bucks should be kept for future breeding. This also allows for the removal of those animals from the herd which are nonproductive. As a rule, does should produce eight litters a year and when they fall to five litters per year it is time to cull. Because rabbits should breed very easily, you should expect to get 70-80% conception rates. If the buck is not producing at this level it is probably best to cull the buck. In either case, be sure that there is not a nutritional, management or disease problem that is contributing to these low conception rates before you cull these animals.

Another idea which is helpful is to keep track of the total number of kits born into a litter and subtract from this the number of kits alive at 21 days of age. This will give you a "nest mortality" figure which may encourage you to improve some management practice. Some breeders like to weigh the entire litter at 21 days of age. The litter that weighs the most is usually indicative of the best mother. Because the buck may sire 10 litters during the same time period that a doe has one litter, his genes are perhaps more important for the overall herd than is the does.

It has been shown that does with heavy fryers (young) at 56 days of age are obviously better milkers. Therefore, fryers are weighed at 56 days of age and this information is also used in selecting breeding stock. The number of kits at 56 days of age can be compared with the number at 21 days of age. Any reduction in numbers is an indication that the manager may have a feed problem, a sanitation-ventilation problem, or a disease problem. The better you can manage your reproduction, nutrition and housing program, the more efficient you will be in producing large numbers of healthy rabbits and at a minimal cost.

SEX DETERMINATION IN YOUNG RABBITS

Hold the rabbit around the chest with your left hand and hold the front legs forward alongside the head. Turn the rabbit over on its back while holding the hind legs with the right hand. Now use the first two fingers of the right hand to depress the tail. Then move the thumb up to carefully depress the sex organs exposing the pink mucous membranes. The buck will have a penis with a rounded tip, while the female will have a membrane that protrudes and forms a slit with a small hole near the anus end (Figure 34). Young rabbits can sometimes be sexed as early as one week, but sexing rabbits at weaning is much more practical.

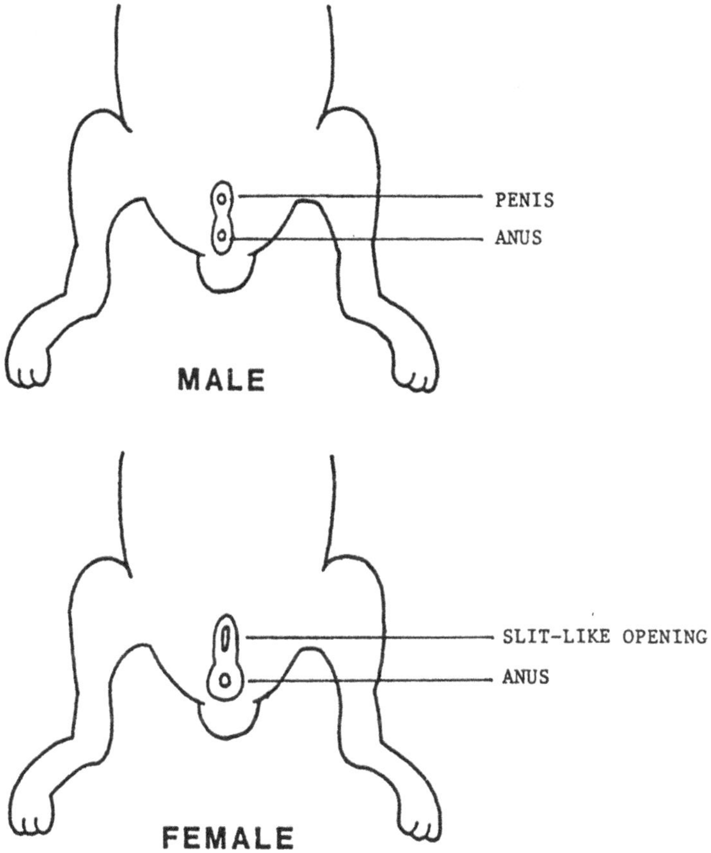

Figure 34: Sexing young rabbits

BREED SELECTION

After the hutch is constructed, the rabbits can be obtained from another farmer or from a commercial breeder. Different breeds can be selected for various characteristics. Mature rabbits of the smaller breeds, such as the Netherlands dwarf, weigh 1.5 kg to 2 kg (3-4 pounds) and large breeds 6 kg to 7 kg (13-16 pounds). The medium and large breeds are most suited for meat and fur production. New Zealand, Californian, Champagne d'Argent, Chinchilla and Flemish Giant are excellent breeds for meat production (Table 2). While the white breeds such as New Zealand and Californian are the most desirable for fur production, the New Zealand white is also the breed used most often for meat. The Californian is the second most common meat breed.

Often rabbit raisers will cross Californian bucks and New Zealand does. One buck can service 2 to 10 does, and the breeds or crossbreeds are not critical for the small producer. The most important objective is to obtain healthy rabbits initially. Rabbits that come from different sources should be housed in separate quarters for 30 days before putting them with other animals in the rabbitry to prevent introducing new diseases into an established herd.

Table 2
SOME COMMON BREEDS OF RABBITS USED FOR MEAT AND FUR PRODUCTION
(Farm Agriculture Handbooks No. 309, USDA)

BREED	DESCRIPTION	WEIGHT (KG)	USES
Chinchilla	Rabbit that resembles the chinchilla in coloration	4.5.5	Fur and meat
Californian	White body with darker nose, ears, feet and tail	3.6-4.7	Fur and meat
Champagne d'Argent	Undercoat dark slate blue, surface coat blue-white or silver, long black guard hairs	4-5.5	Meat
Flemish Giant	Gray, tan, black, white	10+	Meat
New Zealand	White, reddish-brown, black	4-5.5	Fur and meat (white color only)

RABBIT WOOL PRODUCTION

Angora rabbits are raised primarily for their fur or wool. Wool on this variety of rabbit will grow at the rate of about 2.5 cm per month. Usually the wool is trimmed from the rabbit at three month intervals with total production being about 450 grams per year. The wool is valued for its fluffiness, softness and warmth, but usually has to be blended with with other fibers to get a better strength and durability. The Angora rabbit can, of course, be used as a dual purpose animal for both meat and wool production in those areas with cooler temperatures favoring this type of rabbit.

Herd bucks and does are kept in individual hutches, but extra rabbits can be kept in groups or colonies. Castration of bucks in the colony is necessary to prevent unwanted breeding and to help prevent fighting among the bucks.

Because of their thick coats, Angora rabbits should be handled periodically to determine how thin or plump they are, and adjust their feed accordingly. Self-cleaning types of floors for the hutches and colonies are recommended to keep the wool as clean as possible. Also, it is a good idea to shear animals just prior to mating to decrease matting and loss of wool during pregnancy.

Shearing can be accomplished with a hairbrush that has a single row of steel bristles and a pair of barber's scissors. Electric clippers can also be used if available.

Section 5

Nutrition

NUTRITION

INTRODUCTION

Although it has been said that rabbits can be fed any leftover food product, this is only true if specific nutritional levels are met. Different foods contain different levels of those substances which an animal uses to build its body, reproduce and produce energy. These substances are called nutrients, and there are basically six types of nutrients: Carbohydrates, fats, proteins, vitamins, minerals, and water.

Rabbits are herbivorous (plant eating), monogastric (simple stomach) animals with a unique large intestine. The potential of rabbits as meat producing animals is related in part to their ability to utilize fibrous feedstuffs that cannot be effectively consumed by humans. In developing countries, rabbits can be raised to convert available forages and feed by-products into high quality human food.Rabbit meat is high in protein and low in fat, cholesterol, and sodium.

The ability of the rabbit to gain weight on a small amount of feed is very good. Several studies have reported a feed to gain ratio of less than 4 to 1, on a high alfalfa diet. This is much lower than the feed to gain ratio of most other animals on a grain diet.

In the United States of America, the majority of rabbit feed utilized is pelleted. The use of pelleted feed has several advantages, including the reduction of dust, ease of handling, and prevention of animals from sorting through the feed and eating only the more palatable feedstuffs. Rabbits prefer a pelleted form of diet over the unpelleted form when given a choice. They have a higher growth rate compared to rabbits fed the same diet in an unpelleted form. Also there is more feed spillage and wastage with the unpelleted diet.However, in situations where pelleted feed is impractical and/or too expensive, the feeding of unpelleted diets may be utilized with satisfactory results.

The rabbit's intestine is sensitive to many factors, such as pH, starch content and osmolarity. With this in mind, feed changes should be done gradually, over a four-to-five day period in order to allow growth of normal organisms required for digestion.This is especially important in the four-to-twelve-week-old rabbit. During the diet change, old and new foods should be combined, gradually increasing the amount of new diet while decreasing the amount of old diet.

DIETARY REQUIREMENTS

Carbohydrates:

Carbohydrates (starch) are a major source of energy and are important for growing, reproduction, lactation, keeping warm and several other body functions. Food sources which are high in starch include pasture grasses, bananas, cassava, potatoes, hay, and grains such as corn and other cereals. Energy requirements for the rabbit have been estimated to be 2500 kilocalories (Kcal) of digestible energy per kilogram of diet for growth, gestation and lactation. Approximately 2100 Kcal of digestible energy per kilogram of diet is required for maintenance.

One can divide carbohydrates into two main categories: 1) readily available carbohydrates (RAC), and 2) fiber- associated carbohydrates. Unfortunately the rabbit prefers, but is unable to properly digest, large quantities of high energy (RAC) feeds and this has often resulted in two prominent types of enteritis: enterotoxemia and mucoid enteropathy. Both of these diseases are more common with high-energy, low-fiber diets. One easy way to control these diseases is to provide high-fiber, medium-energy feeds. It has been shown that by maintaining 12-15% dietary crude fiber levels, the doe can often be allowed to eat free choice without becoming obese. Of course the amount the doe eats will vary with the stage of lactation or gestation. Because the doe has usually lost weight during heavy lactation, it is important that she be allowed to recover this weight loss as rapidly as possible. The most common cause of difficult breeding, low conception, abortion, fetal resorption, still- births, and small litters is underfeeding. This is especially true for does already maintaining four to five week old litters. Does with litters of seven or more may eat nearly three pounds of a high fiber diet a day.

Indigestible dietary fiber is also important for maximum growth rate and to prevent hair-chewing. Rabbits fed low-fiber, high-energy diets have reduced gains and more pronounced fur-chewing. Fur-chewing may lead to increased problems from gastric trichobezoars (hair balls). Generally, a diet of 16 to 20% crude fiber is recommended for rabbits. Some evidence exists that particle size of the fiber is important (i.e. finely ground alfalfa is less effective than coarsely ground or checked alfalfa). Diarrhea will be seen with dietary fiber levels that are less than 6%. Although fiber is an important component of a rabbit's diet, too much fiber may also cause problems. Digestion of fiber in the rabbit is much lower than in practically any other animal. It has been suggested that high dietary fiber levels, over 22%, may predispose a rabbit to develop fecal impaction.

Fats:

Fats, like carbohydrates, are also a major source of energy in a diet. Because fats are easily digested and highly palatable, rabbits like to eat fatty foods. A dietary fat content of 3% is generally thought to be adequate for rabbits, although dietary fat levels as high as 25% have been fed without adverse effects.

Protein:

Proteins are important in maintenance, growth, reproduction and milk production. These dietary components are found in cereals, grains, by-products of oil crops such as coconuts, soybeans, peanuts and in young green grass, green legumes and hay. Protein requirements for rabbits are 15 to 16% for growth and 18 to 20% for lactation. There is some evidence that bacterial synthesis of protein in the cecum contributes to the amino acid requirements of adult rabbits. It is thought that bacterial synthesis of protein makes only a minor contribution to the amino acid needs of growing rabbits. It is important to remember that the amino acids found in proteins are the building blocks of the body; therefore, protein quality of the diet is important in rabbit nutrition.

Even though 18-20% protein is necessary for optimal growth rate and performance for lactating does, they can also be fed diets containing 16-17% protein. At the lower dietary protein level, there is not a great deal of difference in doe productivity. The economic advantages, particularly in back-yard rabbitries, of the lower protein diet may off set the slightly reduced performance.

Protein deficient diets may result in inadequate milk production and poor growth of newborn rabbits, cannibalism and increased fur-chewing.

Vitamins:

Vitamins are substances that activate all body processes. Domestic rabbits require vitamins A, D, E, K and the B vitamins. Supplementation of the diet with vitamin C is not necessary. Generally speaking, a diet consisting of 30 to 60% good quality dried forage will meet the rabbit's requirements for the fat soluble vitamins (A, D, E and K). Vitamin D is also obtained from the sunlight.

Rabbits produce two kinds of feces, a soft, mucous form at night and a firm pellet during the day. The soft feces are ingested directly from the anus. This practice is called coprophagy. The night feces may consist of 30 to 80% of the total daily excreta. Night feces have a similar protein, minerals and fiber composition as day feces, but have a much higher concentration of niacin, riboflavin, pantothenic acid and vitamin B12. These B vitamins are synthesized by bacteria in the cecum. The practice of coprophagy provides sufficient B vitamins to meet the rabbit's daily requirements.

Minerals:

Minerals are substances which must be presented in the diet for the proper building of bones and teeth. Like vitamins, mineral requirements of rabbits are easily satisfied by a good quality diet consisting of 30-60% dried forage. The major consideration regarding mineral nutrition is an adequate calcium: phosphorus ratio. Generally speaking, a calcium level of 0.40-0.45% and a phosphorus level of 0.22-0.37% will be sufficient for growth and gestation. During lactation rabbits need higher levels, 0.775% calcium and 0.50% phosphorus. Most forages and feeds contain the necessary minerals; however supplements may be needed.

Water:

Clean, cool water should be provided for rabbits at all times. Water helps keep the animal cool in hot weather, but more important is the fact that water is essential for the animal to keep its body tissues alive and healthy. An animal's body is made up of 70% water, thus the need for clean drinking water cannot be over-emphasized.

DIETARY COMPONENTS

Grains:

There is no real health advantage to feeding diets consisting of grains over commercially available pelleted diets. However, if pelleted diets are not practical and/or affordable, grain may be a satisfactory component or supplement to a rabbit's diet. If home grown grains are available and are not needed for either human or other animal consumption, they may be used successfully as a component of the rabbit's diet.

Barley, corn, soybeans, and cottonseed have been successfully fed to rabbits. Other grains that have been fed to rabbits include buckwheat (10% protein) and wheat (soft, 11-13% protein). It has been suggested that one can mix together equal parts of crushed wheat (or oats or corn) and barley, milo, rye, rice or other grain sorghums to feed dry does, young rabbits or herd bucks.

Forages:

If a pelleted diet is unavailable, forages make up the bulk of the daily diet and should be fed free choice. While alfalfa is the most commonly used forage-addition to rabbit diets in the United States, research has also been done on the use of other forage materials. Forages are cheap and abundant year round in tropical countries. A recent study examined the use of woody legumes, non-woody legumes and grasses as dietary additions for the rabbit. Grasses were 6 weeks old and legumes were cut at the flowering stage. Forages were hand- chopped to 1-1.5 cm and wilted overnight and/or steam-pelleted prior to feeding. In general, the woody legumes had the highest nutrient composition, digestibility and palatability. Most forage-based diets need supplementation with grains and other concentrated foodstuffs, commonly prepared as a mash. This is especially true for grasses, as most tropical grasses have very little nutritional value except as a source of indigestible fiber. Note, however than in some parts of the world, local plants fed as forages can supply all the necessary nutrients without the need for supplementation. As a general rule, poor-doing animals, especially during gestation and lactation, require the extra energy and protein found in concentrated diets. It is important to note that the digestibility values of calcium in most grasses, and phosphorus in most forages, are negative, thus requiring supplementation. The following is an alphabetical listing of forages that have been examined as components of rabbit diets.

Alfalfa (Medicago sativa): The protein content of alfalfa deteriorates as the plant grows older. Prebloom alfalfa hay has approximately 19% protein content and early bloom alfalfa hay has approximately 16% protein content. Alfalfa hay, like other legume or grass hays, may be fed free choice in a hay manger or rack on the side of the cage. All hays (and other feeds) should be completely free of molds. Alfalfa is also available in pellet and meal forms. In fryer studies, the use of sun- cured alfalfa meal has yielded better performance when compared to dehydrated alfalfa meal. Sun-cured alfalfa meal is also more palatable than dehydrated alfalfa meal.

Bermuda grass (Cynodon dactylon): Bermuda grass is reportedly toxic to rabbits and therefore, is not recommended as a forage source.

Black locust (Robinia pseudoacacia): Black locust, also called common acacia or false acacia, is a fast-growing, leguminous tree that is readily adaptable to a variety of soils and climates. The young trees can be repeatedly cut so that they grow in a shrubby form from which the leaves are easily harvested. The crude protein level in the leaves is 23%. One study compared the use of dried black locust leaves as the forage component (40% of the diet) of a pelleted diet to the use of alfalfa forage. Average daily gains with black locust leaves were approximately 12 g/day less than those obtained by rabbits fed alfalfa. Average daily gains were still in an acceptable range and no mortality, diarrhea or other adverse effects were reported following the feeding of black locust leaves. Black locust leaves may also be fed fresh to rabbits.

Butterfly pea (Centrosema spp.): The crude protein level of butterfly pea, a tropical forage, is 17.2%. Average daily gains using butterfly pea as 40% of the diet are similar to those seen with alfalfa diets.

Calliandra (Calliandra spp.): This had been used as a forage component of rabbit diets in Indonesia.

Cassava (Manihot utilissima): The crude protein content of cassava is 24.7%. Average daily gains are approximately 10 g/day less than with similar diets utilizing alfalfa as the forage component. The cassava tuber is used as human food and rabbits may be fed with the leaves, a by-product of this human vegetable/crop production. Cassava leaves have been fed to rabbits in Indonesia. The cassava plants grown in Nigeria are reported to have a crude protein of 32% and 18% crude fiber.

Sickle senna (Cassia tora): This tropical forage has a 14.9% crude protein level, when fed to rabbits as 40% of the diet, average daily gains were 5-7 g/day less than those achieved by rabbits ingesting a similar diet of 40% alfalfa.

Comfrey (Symphytum officinalis): Comfrey, a leafy forage plant, may be fed as a fresh green during the growing months or dried and fed as a hay during the winter months. Comfrey is a good protein source, comparable to alfalfa; however, the digestibility of protein in comfrey is only about 70% of the digestibility of alfalfa protein because it is bound to fiber. Comfrey is lower in fiber compared to alfalfa and has about double the ash content. Comfrey also contains allantoin and has been reported to alleviate diarrhea in young rabbits.

Corn plant (Zea mays or Dracaena spp.): Whole corn plant pellets, prepared by harvesting the entire corn plant, dehydrating it and pelleting it, have been fed to rabbits. The pellets are low in protein (6.2], crude protein, and high in fiber.

Gliricidia spp.: This has been used as a forage component of rabbit diets in Indonesia.

Grass hay: Various grass hays such as Prairie, Timothy, Johnson and Sudan may be fed to rabbits. Grass hays contain less protein than legume hays and the diet will probably need an added protein source.

Guinea grass (Panicum maxicum): Guinea grass is a tropical forage with 13.1% crude protein level. Compared with a diet of 40% alfalfa, average daily gains with guinea grass were 5-7 g/day less.

Elephant grass (Pennisetum purpureum): Elephant grass has a 13% crude protein level and a 30% crude fiber level.

Kentucky blue grass (Poa pratensis): Kentucky blue grass clippings can be fed to rabbits at 10% of the diet. Average daily gains are similar to those seen with alfalfa. The grass clippings can be obtained from a sod farm or found in the wild. The young, tender clippings appear to be a good high protein forage for rabbits.

Kudzu: Kudzu is a rapidly growing, coarse, hairy, vine that is native to China and Japan. It is best adapted to drained loam soils of good fertility. Kudzu can be used to prevent soil erosion, as a pasture plant, and also as a hay. Uncontrolled, it may overgrow other vegetation, and can become a pest. Compared to alfalfa, at 10% of the diet, average daily gains were about 8 g/day less, but were still at an acceptable level. Kudzu, as a forage, is palatable and may be especially useful in tropical or warm countries where no alfalfa is available.

Leadtree (Leucaena spp.): This is a tropical leguminous tree producing high protein (greater than 30% crude protein) leaves. Trees can be kept in the shrubby form by repeated cutting. It has been fed successfully to rabbits at levels of 20-30% of the diet, in dried form, with average daily gains of 37 g/day. The digestibility of dry matter and protein is reduced as the Leucaena level in the diet is increased, resulting in poor feed/gain ratios. Urine of rabbits fed Leucaena is deep red to black in color. The origin and significance of the urinary pigments is presently unknown. Leucaena may also be fed fresh, in the green form. It is important to note that Leucaena contains an amino acid, mimosine, which has been toxic for rabbits that were fed Leucaena as the sole forage. In this group of rabbits, various reproductive problems, poor growth and hair loss were seen.

Pawpaw leaves (Asimina triloba): Pawpaw leaves have a 35% 8.8% crude fiber.

Perennial peanut: Florigraze rhizoma is a warm-season, tropical, perennial forage legume, adapted to well-drained soils. The plant is a native of South Africa. When substituted for alfalfa at 40% of the diet, rabbits had similar rates of gain. Some rabbits had diarrhea for the first few weeks on this diet; however, this eventually resolved.

Red clover hay (Trifolium spp.): Red clover hay, approximately 13% protein content, may be used to feed rabbits.

Rice straw (Oryza satire): Rice straw has been fed to rabbits at either 20 or 40% of the diet.Growth rate is approximately 35 g/day.

Sesbania (Trifolium spp.): This is a forage plant that has been utilized to feed rabbits in Indonesia.

Spanish clover: Spanish clover contains 21.4% crude protein. When fed to rabbits at 40% of the diet, the average daily gains are equal to or slightly greater than gains seen with rabbits on a 40% alfalfa diet.

Styloanthes quinesis: This is a tropical forage with 15.8% crude protein. Rabbits fed diets of 40% S. quinesis have average daily gains of approximately 33.5 g/day, which is 7-8 g/day less than gains of rabbits fed 40% alfalfa diet. It should be noted that there was reported a 50% incidence of diarrhea and a 30% mortality rate in rabbits receiving S. quinesis compared with a 20% mortality in the alfalfa diet rabbits.

Sugarcane (Saccharum officinarium): Sugarcane has a 4.8% crude protein level and 21% crude fiber.

Sunflower leaves (Helianthus spp.): Fed as 40% of the diet, sunflower leaves result in poor growth and hair loss problems. Sunflowers contain a phenolic substance, chlorogenic acid, which binds protein and reduces its availability. Hair loss in rabbits fed sunflower leaves is thought to be due to increased fur-chewing caused by an induced protein deficiency. Sunflower leaves appear to have poor potential as a feedstuff for rabbits.

Sweet potato vines (Ipomoea batatas): Have been used to feed rabbits in Indonesia.

Talfairia: Talifairia has a 22% crude protein level and 13% crude fiber.

Talinum: Talinum has a 22% crude protein level and 11% crude fiber.

Vetch (Vinca spp.): Common vetch, approximately 17% protein, may be used as a legume hay in rabbit diets.

Water peanut: In China, water peanut is grown along pond banks. It is collected in the morning, allowed to wilt on mats and then fed later in the day to rabbits. Approximately 70% of the rabbits' diet may consist of water peanut, the remainder of the diet is composed of a variety of weeds and forages.

Wheat straw (Triticum aestivum): Rabbits fed a diet of either 20 or 30% wheat straw had a growth rate of 35-40 g/day.

Winged bean: Winged bean, approximately 18.8% crude protein, fed to rabbits as 40% of the diet results in average daily gains of 10 g/day less than alfalfa. There is a high incidence of diarrhea when rabbits are fed winged bean. It is felt that the diarrhea plus low growth rate may suggest that winged bean contains compounds which interfere with nutrient utilization. Winged bean is a tropical forage. The bean pod has been used as human food, thus rabbits can be fed a by-product of vegetable/crop production.

Seeds:

The following is an alphabetical list of seeds that have been examined as components of rabbit diets.

Amaranthus: Seed heads or both black and white species of amaranthus have been fed to rabbits. Both types resulted in poor average daily gains. Feeding of seed heads from the black-seeded amaranthus resulted in an 80% mortality rate due to a possible hepatotoxin.

Barley screening (Hordeum vulagare): Rabbits fed a diet of 20% barley screenings, which are a by-product of processing plants, as a fiber source, have average daily gains equal or superior to rabbits fed a similar diet with 22.5% dehydrated alfalfa.

Canola meal: Canola meal or low glucosinolate rapeseed meal is a good replacement for soybean in rabbit diets.

Cottonseed (Gossypium spp.): Rabbits fed a diet of 43% cottonseed have no significant differences in gain compared with rabbits fed a soybean meal diet. Cottonseed meal, prepress solvent extracted, has been fed to fryers as 4, 8.2, 10 and 16.5% of the diet. No adverse effects attributed to the cottonseed meal were noted as assessed by rate of gain, mortality, or feed efficiency. No effects on the semen of bucks fed 20% cottonseed meal for 150 days were noted. At these levels cottonseed does not appear to have adverse effects on male rabbit fertility. Long term studies on does need to be done.

Flax (Linum usitatissimum): A diet of 30% flax has been fed to rabbits. No significant differences in gains are noted between rabbits fed flax seed and a soybean meal diet.

Lupin grain: Diets of 7, 14, and 21% lupin grain, 40% crude protein, have been fed to rabbits. No significant differences in average daily gain or feed intake are seen in rabbits fed lupin grain as compared to soybean diet.

Maize (corn) bran: Maize bran has 11.2% crude protein and 7.8% crude fiber.

Peas (Pisum sativum): Peas, 26-27% crude protein, have been fed to rabbits as 30% of the diet. No significant differences in average daily gain or feed intake are seen in rabbits fed peas and those fed soybean meal.

Rice and wheat bran: Rice and wheat bran are excellent sources of protein and energy for rabbits. Wheat bran contains approximately 16% protein and 10% fiber plus vitamin E.

Safflower (Carthanus tinctorius): Rabbits fed a 34% safflower diet have no significant differences in gain compared to rabbits fed soybean meal.

Soybeans (Glycine spp.): Soybean meal contains approximately 45 to 48% protein. It may be added to rabbit rations at a percentage of 5 to 7% of the total ration. Adding too much soybean meal can cause digestive problems in rabbits, especially young rabbits.

Soybean hulls: Rabbits fed a 10% soybean hull diet as a fiber source have average daily gains that are statistically equal or superior to a diet of dehydrated alfalfa.

MISCELLANEOUS CATEGORY

The following is an alphabetical list of various supplements and components of rabbit diets.

Banana (Musa paradisiaca var. sapientum) and plantain leaves (Plantago spp.): Mixtures of banana and plantain leaves have been fed to weanling rabbits. The leaves, comprising 30% of the diet, were fed either fresh, dry or half and half. No difference in performances was seen between fresh or dry leaves with average daily gains of about 20 g. Banana leaves are 20% crude protein and 24% crude fiber. Plantain leaves are 18% crude protein and 20% crude fiber.

Beet pulp (Beta vulgaris): Rabbits fed a 15% beet pulp diet, as a fiber source, have average daily grains that are statistically equal or superior to those of rabbits that are fed dehydrated alfalfa.

Blood meal: Blood meal has 24.3% crude protein and 24% crude fiber.

Brewers grains: Dried brewers grains contain 28% crude fiber.

Brewers yeast: Dried brewers yeast contain 46.3% protein and 12% crude fiber.

Cabbage (Brassica oleracea var. capitata): Fresh cabbage leaves have been fed to weanling rabbits in which 5, 15, and 25% levels replaced corn in the diet. Average daily gains were 13.2, 18.1 and 12.4 g/day respectively.

Citrus pulp: Dehydrated lemon and orange citrus pulps have been fed to weaning rabbits as 20% of the diet, replacing 20% corn in the diet. Digestibility of the citrus pulp is similar to the corn diet. Gains with the citrus pulp diet were about 20 g/day compared to 24 g/day with corn.

Corn husks: Corn husk contains 2.6% crude protein and 31.6% crude fiber.

Cottonseed cake: Cottonseed cake contains 22.4% crude protein and 25% crude fiber.

Copper: Conflicting literature reports exist on whether or not the addition of copper to rabbit diets is beneficial. One study examined the effects of 100, 150 and 200 ppm copper sulfate in diets that were low (14), medium (18), and high (22) in protein. This study found that growth was significantly enhanced at 200 ppm copper sulfate for the low and medium protein levels, but not the high protein level diet. This suggests that as dietary protein levels increase, the effects of adding copper to the diet decreases. The mechanism of action is unknown; however, cupric ions activate pepsin, an enzyme involved in protein utilization, which may be part of the reason for the growth promoting effect of supplemental copper. Another study did not show any effect on weight gains when copper (250 ppm) was added to the diet. Some studies have also suggested that dietary supplementation with copper has some protective effect against diarrhea and mortality in 4 to 10 week old rabbits. Other studies have not supported this finding.

Goat's milk: One study gave rabbits free choice pellets and goat's milk. Rabbits given goat's milk consumed less than half the amount of pellets eaten by those control rabbits receiving water and pellets ad lib. When excess milk is available, the cost of feeding rabbits could be reduced by giving milk instead of water.

Palm kernel meal (Arecaceae family): Palm kernel meal contains 21% crude protein and 10% crude fiber.

Peanut cake (Arachis hypogaea): Peanut cake contains 52% crude protein and 5% crude fiber.

Peanut hulls: Peanut hulls contain 5.7% crude protein and 31.8% crude fiber.

Poultry broiler finisher pellets: Poultry broiler finisher pellets (18% crude protein) can be fed in limited quantities along with free choice grass to produce rabbit meat at a cost comparable or less than that of fresh chicken and pork. A study in Trinidad fed each doe 80 g/day for day 1-23 of gestation and 100 g/day the least 8 days of gestation. The schedule for concentration feeding from kindling to weaning at 28 days was: 120 g/day first 7 days then 120 g + 10 g/kit/day for the next 21 days. Fryers were offered 30 g/day the first week post weaning and this amount was increased by 10 g every 7 days, until the animal was 9 weeks old. Animals 9 weeks of age and older received 80 g/day of the concentrate.

Poultry manure: Dried poultry manure contains 26% crude protein and 18.6% crude fiber.

Rabbit manure: Dried rabbit manure contains 36.7% crude protein and was fed in one study as 10-15% of the total diet. Average daily gains of rabbits fed manure were similar to those fed alfalfa. Rabbits fed manure had a decreased incidence of enteritis. If rabbit manure is treated with alkali, the crude protein level is 0.4% and the digestible content is 1970 Kcal/kg (vs. 722 Kcal/kg pre-treatment) which is similar to alfalfa.

Rice husks: Rice husks contain 5.2% crude protein and 43.2% crude protein.

Urea: Urea has potential as an inexpensive substitute for low protein in rabbit diets. Experimental studies have reported somewhat variable results. Early studies, using weanling rabbits, showed no beneficial effects of the addition of urea to a low protein diet. More recent studies have shown that a urea diet does not support body weight gain and nitrogen retention as well as soybean meal diet in adult animals; however, fryers appear to utilize urea more efficiently than adult animals. With fryers, gain is actually slightly superior with a urea diet as compared to a soybean meal diet.

Vegetable leaves and tops: Various vegetables and vegetable leaves have been fed successfully to rabbits. These include collard leaves, carrots and carrot tops, dandelions, potatoes and potato peels, turnips, sweet potatoes, parsnips, English pea and green bean vines. Raw potato peels are high in Vitamin B_6.

Zeolite: Zeolite is a buffering compound, crystaline alumino- silicate, with ion exchange properties. Addition of zeolite to the diet in either 1.5% or 3% has results in significant decreases in mortality due to enteritis.

TYPES OF DIETS

Although many individual food products have been discussed in this section, there may be a commercial supplier of concentrated feeds in the area. If rabbit chow is not available, one may substitute a pelleted pig grower ration or even a pelleted chicken feed. It is especially helpful to use the commercial products as a supplement to the natural foods during the rapid growth periods of the young, during the last ten days of pregnancy, and during lactation. Some recommended rates for using the commercial supplements for New Zealand White rabbits are: 90 g per day for growing young; 90 g per day during pregnancy; 150 g per day during first 3-4 weeks of lactation; add 10 g more per kg during the last few weeks of lactation (1 ounce = 28.25 grams).

Generally one diet is used for the entire herd. However, the protein and energy requirements for lactation are higher than for growth. If only one diet is used for all rabbits, for example, a diet suitable for a lactating rabbit, then fryers will receive excess protein and the high energy level may induce diarrhea. If a diet adequate for fryers is fed to all rabbits then one may see decreased productivity in does.

It has also been suggested that at least two diets could be advantageously used for fryers. One diet would be a high fiber diet weaning diet and the second diet would be a high energy finishing diet. More research needs to be done in order to develop optimal diets for various production phases.

Section 6

Diseases

INTRODUCTION

Obviously, one of the most devastating and discouraging aspects of rabbit production is disease. The term disease is used here in the broad sense of various illnesses, disorders or ailments, which may cause poor performance, sickness, other abnormalities, or death. Disease states may result from infection by microorganisms and parasites (infectious diseases) or from injury, poison, and functional or metabolic disorders (noninfectious diseases). Rabbits may be affected by both classes, but the infectious diseases are more common and may cause high mortalities.

The success or failure of a rabbit operation is probably dependent upon the owner's ability to keep disease at a minimum. Most diseases occur because of management mistakes or lack of knowledge. Mortality in many herds averages 20% or more of all animals born. The total loss or rabbits due to disease in the United States averages 20-25%; this includes kits born dead, nest box deaths, fryer deaths, and the loss of does and bucks. In other areas of the world it seems reasonable to expect that death losses would be at least this high. This section will discuss bacterial, viral, fungal, parasitic, hereditary and miscellaneous diseases often seen in the rabbitry.

When well fed and properly housed in draft-free sanitary hutches, rabbits will live relatively free of diseases. The best cure is always prevention. Because medications are not always available, the following paragraphs will discuss natural animal medicines and their reported uses in preventing and curing common rabbit ailments.

Many wild animals have an innate ability to choose what is needed to get well. Old-time rabbit breeders have used this information to discover natural medications with which animals in the wild treat themselves.

Feeding the leaves of plantain, dandelion, or the leaves and bark from elm trees once or twice a week may help keep a rabbit herd from developing the need for more drastic chemical drugs. These natural materials appear to be effective in preventing common scours (diarrhea) that sometimes afflict the very young rabbit.

The bark and the limbs of the sassafras tree also appears to relieve digestive upsets in newborn rabbits. Once the bark is all eaten from the piece of sassafras, it is not uncommon to see the doe place the wood in her water bowl allowing the wood to seep into the water making tea. Breeders have felt that either the bark or the tea is helpful in preventing diarrhea.

It has also been shown that dry leaves of certain trees may have about the same effect as green leaves. The dry leaves may be gathered when they are fresh, clean and free of mold, decay and insects. Rabbits seem to prefer the leaves of oak, elm, sassafras and willow trees. Apparently the bark of the willow tree also contains substances that are beneficial to rabbits.

Blackberry leaves have long been considered an excellent cure for scours. Because the rabbit is highly susceptible to chemical toxicities, it is very important to never use the leaves, bark, grass clippings or berries that have been sprayed with insecticides or herbicides.

Another natural remedy for diarrhea or scours in rabbit herds has been the use of apple cider vinegar. Apparently the apple cider contains large amounts of potassium, which may be the major beneficial ingredients for rabbits. The apple cider vinegar is diluted to 5% acidity, and 30-60 mLs are added to 3.5L of fresh clean drinking water. The vinegar treated water is then given daily over a period of two to three months. Some breeders have claimed that giving rabbits this vinegar-water daily will also control coccidiosis. Other breeders have actually reported that the apple cider vinegar in the drinking water will increase the reproductive rate of rabbits in a rabbitry.

Apple cider vinegar has also been shown to improve the rabbits' coats, giving them a healthier sheen even in hot months when rabbits are likely to be shedding at least some of their heavy fur. Improvement in rabbit coats is probably due to the acidity of the vinegar, which can effect the pH balance of the skin and give the hair follicles a healthier base in which to grow.

A natural remedy which has been used to combat a lack of protein in the diet is feeding comfrey (Symphytum officinalis). Comfrey contains most of the amino acids that are the building blocks of protein, plus it contains the curative substance allantoin which is commonly used in many medications. Comfrey is fed as a fresh green plant during growing months or it may be dried and fed during the winter months. Like alfalfa, it should be fed as fresh as possible to have the greatest effect, but drying the summer crop for winter use is as practical as using other hays. It offers more extensive benefits with regular use. Many rabbit breeders have reported that their rabbits are very seldom sick with colds, numerous rabbits suffering from various types of diarrhea have cleared up, and those suffering from run down conditions due to a heavy breeding schedule and nursing, have all been shown to improve or disappear after the feeding of comfrey. Recent studies have indicated that comfrey may have some carcinogenic (cancer-causing) activity; therefore, it is recommended that this food source be used with discretion.

In raising rabbits you may come in contact with some unusual methods for treating diseases. One such medication is pekoe tea given in the drinking water. This substance has been shown to be effective against some bacterial diseases such as Tyzzer's disease. One large tea bag is diluted in 4L of water, cooled and given back to sick rabbits as drinking water. The tannic acid in the tea apparently is effective in killing many bacterial diseases. The weak tea has also been shown to clear up cases of ordinary diarrhea when given with clean, bright straw to eat.

As in many other animals, stress is known to produce or precipitate diseases in rabbits. Common stresses include such things as extreme heat in the summer, cold in the winter, and noises of various kinds. Perhaps the greatest stress placed upon the rabbit is lack of proper sanitation. Living in filthy quarters can stress any animal, including those much larger and stronger than the rabbit. Accumulation of manure and urine create an ammonia gas in the building and hutch. This in turn causes a severely oppressed air quality which has its effect on the respiratory system of the rabbit. As with heat, those carrying or nursing young and little rabbits are the most susceptible. When the rabbitry operator can smell ammonia gas in the rabbitry, damage has already been inflicted on the lungs, nasal passages, and bronchial tubes of the rabbit.

DISEASE DIAGNOSIS

As a rabbit raiser, you should become very familiar with the normal appearance and physiological characteristic of a rabbit. Refer to Table 2 for physiological parameters. Rabbits suffering from a disease usually exhibit some characteristics which make them appear or act differently from healthy animals. The following clinical signs are indications of a disease or abnormality, and each should be evaluated carefully:

1. Failure to eat;
2. Abnormal breathing;
3. Discharges from the eye and nose;
4. Loss of weight;
5. Diarrhea;
6. Listlessness;
7. Blood in excretions;
8. Skin lesions;
9. Fever;
10. Increased pulse rate;
11. Increased respiratory rate;
12. Change in color of mucous membranes.

Close observation of clinical signs while the rabbit is living may allow for diagnosis. Other diseases cannot be diagnosed without a necropsy examination, and in some cases, special techniques. A necropsy (post-death examination) should be performed by a veterinarian as soon as possible after death, since tissues degenerate rapidly and examinations several hours after death are of little value unless the carcasses are preserved. Refrigeration can be used to delay tissue decomposition if necropsies must be performed at a later date.

COMMON DISEASES

The following diseases are grouped as: (a) miscellaneous and noninfectious diseases and ailments, (b) hereditary diseases, (c) viral diseases, (d) bacterial diseases, (e) fungal diseases and (f) parasitic diseases—external and internal parasites.

MISCELLANEOUS AND NONINFECTIOUS DISEASES

Cannibalism:

Most cases of cannibalism, which is the act of one rabbit eating another rabbit, are the result of the diet being inadequate in either quality or quantity, or because the does are disturbed following kindling. In a breeding program it is essential that the doe be provided with adequate housing and diet in order to prevent this condition. Does which continue to destroy their litters should be removed from the breeding population.

Heat prostration:

In tropical climates, care should be taken to prevent rabbits from getting too hot and collapsing. Adult animals suffering from heat may be relieved by spraying them with water or placing a wet burlap feed sack or other cloth on the cage floor for the animals to rest on. Bedding and fur should be removed from the nest box to allow free circulation of air in boxes with infants. Clinical signs of heat prostration are increased respiratory rate, depression, and evidence of a bloody discharge from the mouth and the nose. Mortality rates can be high unless environment temperatures can be reduced below 33°C (92°F). Pregnant does are most often affected, and young rabbits in the nest box are particularly susceptible when there is excessive bedding and little ventilation.

Perhaps the most important consideration is preventing additional rabbits in the colony from suffering from heat prostration (heat stress). Measures should be taken to cool the ambient air around the hutches as is described in Section 2, "Care of Rabbits During Extreme Heat".

Broken back:

Spinal fracture is an injury which may occur if the rabbit kicks excessively when the rear legs are not properly supported. Paralysis may extend from the middle of the back resulting in an animal that cannot use its back legs. The animal moves with its front legs and drags the hindquarters. Often times the urinary bladder may become greatly enlarged due to the difficulty in voiding urine, giving the animal a bloated appearance.

Other conditions which may produce spinal injury include using a cage short for the size of the animal or injuries which may occur at night when predator animals invade the rabbitry. A rabbit with a dislocated spine usually cannot be treated and should be destroyed. Sometimes, if the spinal cord has not been too severely damaged partial recovery may occur.

Sore hocks:

Although the scientific name for this disease is "pododermatitis", the name "sore hocks" is much more descriptive of the inflamed bare spots, devoid of fur, which are found on the bottom surface of the hind legs. Severe cases will develop secondary infections with Staphylococcal organisms producing abscesses. If a rabbit walks with an unusual motion or the rabbit sits with more weight on the front feet than normal, the disease sore hocks should be suspected. Although both front and hind feet may become involved, the animal usually throws more weight to the front feet due to the painful hocks of the infected hind feet.

A genetic predisposition for this disease occurs in certain strains of rabbits. If the animal is inherently nervous, excitable, and is active with excessive foot stomping, this may contribute to the development of sore hocks. Also, larger breeds and fat animals seem to have a higher incidence compared to smaller breeds. Since there is an inherit predisposition to this disease, rabbits with a habitual tendency towards sore hocks should not be used as breeders. Environmental factors which may be involved in this disease include the type of wire on the cage floor, sharp edges, the cleanliness of the cage floor, and the condition of the cage floor. Filthy caging and improper housing can easily be corrected by proper management.

The use of ointments, sprays, and many other medications will produce temporary improvement in rabbits suffering with sore hocks. Placing the rabbit on a soft flooring such as sawdust may also help the feet to heal. Unfortunately when the rabbit is returned to its original housing or his treatment regime is discontinued, the condition will usually reappear. Most rabbitries believe it is more economical to cull (remove) the rabbit, and avoid using this animal for breeding stock.

Malocclusion:

Malocclusion and tooth overgrowth result when (for genetic, dietary, infectious, or traumatic reasons) open rooted teeth do not meet properly and are therefore not worn down as they would normally. In the rabbit, the teeth commonly involved are the incisors, which, if not constantly worn down, will grow approximately 10 cm (4 inches) per year. Malocclusion in the rabbit is probably an autosomal recessive (inherited) trait involving an abnormally short upper jaw. Buck teeth are also an inherited characteristic and this undesirable gene will increase in a breeding colony unless specific culling is carried out to eliminate these animals. There is no cure for this condition other than culling. Temporary treatment involves cutting away the excessive growth using heavy scissors or wire cutters. This will allow the young to eat and obtain slaughter weight. The teeth will continue to grow during this period and repeated trimming may be required every two to three weeks.

Hairballs (Trichobezoars):

Many rabbits ingest small amounts of hair during the normal practice of grooming. Several rabbits in a hutch may eat body fur, eyelashes, and whiskers. Accumulation of hair within the stomach is thought to be attributed to the inability of the rabbit to vomit and the small pyloric lumen. In either case the fur mixes with the stomach contents and forms a firm ball which cannot pass through the pyloric valve between the stomach and the small intestine. Causes for this condition include hair loss; lack of dietary fiber; protein, copper, or a magnesium deficiency; abnormal grooming behavior; and excessive shedding of a naturally long hair coat.

Clinical signs for hairballs include anorexia (failure to eat), weight loss, agalactia (poor milk production), depression, absence of feces and eventually death in three to four weeks from starvation and metabolic abnormalities. Diagnosis is often difficult, but usually can be made using palpation and/or radiography.

Although routine treatment with mineral oil (10mL at a time) on a monthly basis may prevent hairballs, the formation of a ball too large to pass out of the stomach cannot be treated by this method. A wetting agent, dioctyl sodium sulfosuccinate 5% in propylene glycol or one of several flavored laxatives and lubricate preparations, can also be used. Of recent interest is the use of a digestive enzyme (papain) and/or pineapple juice, which may break down the fiber particles and allow for passage of the hairball. Surgical removal of the hairball can be used as a last resort. Postoperative care should include fluids, antibiotics, and hay or calf manna. Unfortunately, recurrences are common. Even though there is no effective method of treatment, many breeders feel that a 15% fiber level is probably the minimum that should be fed to either growing or producing rabbits. Apparently the increased fiber content in the diet helps prevent the formation of hairballs.

Coprophagy:

This is a normal characteristic of the rabbit in which it actually eats its own fecal pellets, and this usually occurs at night. Fermentation of feces in the large intestine supplies an abundance of certain B vitamins to the fecal pellets, probably improves the quality of the protein in the soft pellets, and improves fiber breakdown by bacterial action. The rabbit actually takes these soft pellets from the vent as they are passed and reingests them. It is thought that this second passage of food through the digestive tract allows for increased nutritive values to be derived from the food. Coprophagy can occur during the day, although it is usually a night practice and should not be interpreted as a disease or an indication of a depraved appetite.

Yellow Fat:

Some rabbits develop a very dark yellow deposition of fat. This is a genetic trait determined by a recessive gene which prevents the reduction of the xanthophyll pigment to a colorless product. This is not a disease condition; however, it is often an objectionable trait because consumers prefer white fat in meat rabbits. The only solution is to cull those breeders producing this yellow fat syndrome.

Wet Dewlap:

The dewlap is a loose fold of skin under the neck, which is very large in female rabbits. During warm weather, rabbits will tend to drink large amounts of water and this may predispose this area to excessive moisture. If this dewlap remains damp, the fur will discolor and may develop a secondary bacterial infection. Usually the hair will turn green (from bacterial pigments) and the skin will become inflamed beneath the matted fur.

Treatment involves clipping away the matted fur and treatment of the skin with an antibiotic ointment. The removal of water crocks, fixing leaking sipper tubes and, in general, providing for a dry environment will usually prevent this condition. "Toys" (extra bowls or other objects) in the cage are also helpful to prevent the rabbit from "playing" in the water.

Hutch Burn:

The reproductive organs and teats are very sensitive to urine and fecal contamination. These tissues are easily irritated and may have brownish crusts covering the area with a bloody exudate. This condition is often confused with rabbit syphilis or vent disease. Treatment involves warm compresses, the application of camphor oil and massage to remove the milk. If the teat and its milk gland are severely caked it may be necessary to lance it and thoroughly irrigate the tissues with clean water. Topical and systemic antibiotics are also recommended. Often times this condition will predispose the female to subsequent mastitis, which is a severe disease of the milk glands. The easiest prevention is to provide a clean dry environment.

HEREDITARY DISEASES

Buphthalmia (Congenital or Infantile Glaucoma):

This eye condition initially appears as an increase in the size of the eye with a gradual cloudiness and a bluish color on the surface of the eye. The eye soon becomes more milky in color, enlarges, and contains increased blood supply to the surface. Eventually the eye is so opaque that blindness results. Because this is thought to be inherited, it is best to remove these animals from the breeding colony.

Hydrocephalus:

This is almost always a condition which is first noticed as an enlargement of the head at birth and then quickly worsens as the animal grows. Usually the young die within a short period of time. Although many reports describe this as an inherited trait, it has also been shown that a lack of vitamin A in the diet of the pregnant doe may be correlated with hydrocephalus.

Splay leg:

This condition is similar to the hip dysplasia found in certain breeds of dogs and it is thought to be an inherited condition. Affected rabbits cannot move the involved limb(s) toward the body resulting in a "splay leg" appearance for the animal. Again, because of the potential genetic involvement, all affected animals should be removed from the breeding program.

VIRAL DISEASES

Rabbit pox:

Rabbit pox is a severe, highly fatal disease that can occur with or without clinical lesions. It is caused by a pox virus which is closely related to the vaccinia virus and it is transmitted through nasal secretions.

Clinically, animals are febrile with a profuse watery diarrhea, enlarged inguinal and popliteal lymph nodes and cutaneous skin lesions. The skin lesions are usually seen as a rash covered by a crusty exudate. Inflammation of the eyes and surrounding structures (blepharitis, keratitis, and purulent conjunctivitis) is often present. Mortality is highest in the unweaned young where it may approach 75% in epizootics (sudden outbreaks).

Vaccination with vaccinia virus is recommended for susceptible colonies in the face of an outbreak.

Myxomatosis:

This is an infection of cottontails (Sylvilagus spp.) which can easily enter rabbitry colonies by means of mosquito transmission. Myxomatosis is also called **Big head Disease** because of the swelling produced around the eyes, ears, lips and nose. Clinically there is swelling around all body orifices and the ears may become so edematous that they become heavy and pendulous. Diagnosis is usually made on the basis of the clinical signs described and over 80% of the animals will die in 10-12 days. Histological examination of the affected tissues is used to confirm this diagnosis. Some attempts have been made to vaccinate infected colonies with an attenuated rabbit fibroma virus or a modified-live myxomatosis virus.

Rabbit Papilloma:

This is a DNA virus belonging to the Papilloma virus group which usually causes papilloma-type lesions in the cottontail, but has also been isolated from domesticated rabbits. The most common sites for these well keratinized papillomas are the eyes and eyelids.

As these wart-like lesions increase in size they are easily scratched or knocked off leaving an open wound which heals without complications.

It is thought that transmission from the wild cottontail to the domestic rabbits is accomplished via insects such as mosquitos and ticks. The obvious measure is to protect the animals from these arthropod vectors.

BACTERIAL DISEASES

Pasteurellosis:

One of the most serious problems observed in rabbitries is pasteurellosis caused by Pasteurella multocida. This disease results in economic losses through deaths, failure to gain weight, removal of sick animals, and in some cases failure of affected rabbits to reproduce. Rabbits are more susceptible to infection with Pasteurella multocida than most other laboratory animals, and several clinical forms of the disease occur. This disease can manifest itself in a number of different ways, such as metritis (infection of the uterus), orchitis (infection of the testicles), mastitis (infection of the mammary glands), conjunctivitis (infection of the eye), sinusitis (infection of the sinuses), and subcutaneous abscesses. Of all these conditions, sinusitis (snuffles), is by far the most common. Severe snuffle problems have forced many rabbitries out of the business and cut the profits of most large rabbitries.

Snuffles is most often seen in rabbitries with poor ventilation. The rabbitry is usually over crowded and excessive manure build-up under the cages is common. The odor of ammonia from the breakdown of rabbit urine is often detected. All of these factors put a considerable amount of stress on the rabbit's respiratory system. The mucous membranes of the nasal sinus become infected by bacteria in the inspired air or by direct contact with infected animals or contaminated objects. The clinical disease is characterized by nasal discharge (mucus or pus). Infection of the paranasal sinuses probably results in a lifelong infection in most cases. Infection may be clinically silent for varying periods of time with intermittent periods of mucopurulent nasal discharge (snotty nose). The discharge may be wiped from the nose with the medial (inner) aspect of the fore feet resulting in a wetting and matting of fur on the feet.

Snuffles is extremely contagious to other animals in the herd. The other forms of pasteurellosis mentioned previously generally begin appearing after snuffles is noticed. The devastating aspect of Pasteurella multocida is that it has been shown to spread throughout the body by several routes including: a) the nasolacrimal duct to the conjunctival sac; b) via the trachea to the lungs; c) by the eustachian tube to the middle ear and subsequently to the inner ear, meninges, and brain; d) via blood to the organs and tissues throughout the body; and e) by lymphatics to lymph nodes and the blood stream. Rabbit mortality surveys have shown that the greatest single cause of death in mature animals is pneumonia following an upper respiratory disease with snuffles. Signs of pneumonia are depression, labored breathing, bluish eye color in albinos, and a nasal discharge. The body temperature is usually above normal (fever).

Many times infected females will not reproduce and are therefore culled from the herd and slaughtered. Necropsy usually reveals an infected uterus (pyometra) which can be traced back to a single buck with a chronic infection of testicles (orchitis). The testicle becomes enlarged and contains an abscess. Pus can be seen when the testicle is cut open. As described above, the infection is transmitted to does by infected bucks during breeding.

Pasteurella multocida is the causative organism found in the majority of abscesses seen in rabbits. Abscesses are usually seen in the subcutaneous areas and occur when the pasteurella organism invades a break in the skin due to a scratch, cut, or sore. Abscesses can also occur internally. Although abscesses can be lanced, drained and treated with a penicillin-streptomycin antibiotic, it is probably best to cull such rabbits from the herd immediately. Treated rabbits often show a recurrence of the problem and may well act as carriers of the pasteurella organism.

Middle ear infection of one or both ears causes filling of the tympanic cavity with a purulent exudate. If the process spreads to the inner ear, the equilibrium of the animal is disturbed and a head tilt or wryneck results. Again treatment is not effective and rabbits with this condition should be culled from the herd. There is nothing wrong with the meat from these rabbits; they can be safely eaten.

Conjunctivitis or weepy eyes is a condition which is often seen in rabbits in the nest box and sometimes in older does. It can be caused by several different bacteria; however, the usual cause is Pasteurella multocida. Infection of the eye usually occurs as a result of extension of the infection of the nasal cavity (snuffles). Antibiotic ophthalmic ointment containing penicillin or chloromycetin are particularly effective in treating this disease (Table 3). If the infection persists or returns, a blocked tear duct can be anticipated which will require a veterinarian's care to correct.

When the organisms enter the blood stream, an overwhelming **septicemia** may develop. Tissue changes usually occur very rapidly in the heart, spleen, and upper digestive tract resulting in the death of the animal within a very short time. A lack of clinical signs and the short duration do not allow time for suitable treatment of this condition.

Because nearly all of the pasteurella isolates are sensitive to penicillin, individual cases can be treated with a daily injection of procaine penicillin 60,000 IU/kg of body weight for ten days. Although it is usually best to cull sick animals from the herd, in some instances it may be necessary to treat the condition on a herd basis. The addition of a broad-spectrum antibiotic to the water such as 300 mg/liter of tetracycline or the use of a feed additive sulfaquinoxaline, at 225 gm/ton of feed or furazolidone at 50 gm/ton of feed, may produce beneficial results. Other antibiotics which have been used include: Sulfamethazine, sulfamerazine, ampicillin, cephalothin, gentamicin and chloromycetin (Table 3). Recurrence of clinical signs frequently accompanies cessation of treatment. It is **recommended** that treated animals **should not** be used for human consumption for a period of at least 72 hours following treatment.

Because pasteurellosis is very difficult to treat and exists in a carrier state, the success of a rabbitry is dependent upon the prevention of pasteurellosis from entering the colony. A good preventive medicine program will rely on the following: 1) purchasing of breeding and replacement stock from healthy suppliers which are free of pasteurella, 2) providing a clean environment with adequate sanitation and the prevention of fecal contamination and ammonia buildup in the environment, and 3) purchasing pasteurella-resistant rabbits (when the premises become severely contaminated). Experimental work is now being conducted to develop a streptomycin- dependent live Pasteurella multocida vaccine for the prevention of rabbit pasteurellosis. Further testing is necessary to determine the value of this vaccine.

Tyzzer's Disease:

This disease is found in mice, rats, hamsters, gerbils, rabbits, guinea pigs, horses, cats, dogs and several wildlife species. Bacillus piliformis is the causative organism.

Transmission is usually by the fecal-oral route, although in-utero transmission has been demonstrated experimentally. A big problem with this disease is that infectious spore-like bodies may survive a year or more in bedding, soil, or contaminated feed. Some form of stress, such as overcrowding or extremes in temperature, is usually needed to precipitate this disease.

The clinical signs of Tyzzer's disease can be either acute or chronic. Signs of the acute form in weaning or stressed animals are diarrhea, listlessness, anorexia, dehydration and death within 48 to 72 hours. Subclinical infections are very common resulting in carriers within a colony which continue to shed organisms. Chronically infected animals exhibit weight loss, rough hair coat and eventually death.

Pathological lesions include necrosis in the wall of the cecum and focal necrosis in the liver and heart. Acute deaths result in edema, congestion, hemorrhage and focal ulceration of the intestine. The intestine is often atonic and filled with a yellowish fluid. The hepatic, intestinal and myocardial foci, which are so characteristic for this disease, probably arise via an embolic shower of organisms from a primary infection in the intestine.

Diagnosis is usually made on the basis of clinical signs and demonstration of the organisms in liver cells. Many times these organisms will give the appearance of a "pile of sticks". A complement fixation test has been effective in diagnosing this disease in Japan.

Treatment is usually not effective as the acute cases die before a therapeutic program can be initiated.

Oxytetracycline in the drinking water at 0.1 g/L for 30 days has been effective in suppressing outbreaks (Table 3).

Tularemia:

This is an infectious disease of wild animals and humans that is caused by Francisella tularensis. Tularemia is rare in domesticated rabbits unless they are exposed to wild animals, birds, rodents, deer flies and ticks. Transmission can be by direct contact, aerosol, or biting arthropods. Organisms can be found in the tissues, blood and feces of infected animals.

Tularemia produces an acute septicemia which usually results in death within one week. Necropsy lesions include pulmonary congestion and numerous small white necrotic foci in the liver and spleen. Diagnosis is based on these necropsy findings and isolation of the organism. This agent is extremely hazardous to humans and must be handled with extreme caution in the laboratory. Treatment of an affected rabbit is not recommended due to the public health hazard. All infected animals should be euthanatized and their carcasses burned.

Spirochetosis:

This disease is caused by Treponema cuniculi. It has been erroneously called "rabbit syphilis", although this disease is not the same as "human syphilis".

The accidental use of infected bucks will quickly spread this disease through a colony. Lesions involve crusty, dry, scaly or edematous sores on the nose, mouth, ears and genitalia. Lesions around the vent easily become irritated from urine and poor sanitation and have resulted in the term "Vent disease". The organism is sensitive to arsenicals and penicillin. A single intramuscular injection of 100,000 units of penicillin is the recommended treatment (Table 3).

Listeriosis:

This disease is caused by Listeria monocytogenes. It exhibits three clinical forms: 1) a blood-born infection in young animals, 2) a brain infection in adults, and 3) a uterine infection with fetal mortality in pregnant does.

Clinical signs include weight loss, incoordination, convulsions, nasal discharge, uterine discharge and abortion. The most consistent lesion at necropsy is liver necrosis which produces small pinpoint necrotic foci resembling Tyzzer's disease and tularemia. Pregnant does and does which have recently kindled are the most susceptible. Diagnosis is usually made at necropsy and so treatment is of little value. If a diagnosis is made, sick animals may be treated with tetracycline or penicillin. Females recovering from this disease are often sterile (unable to become pregnant). This disease is transmissible to humans and a positive diagnosis should be reported to the local health authority.

Enterotoxemia, Rabbit Diarrhea Complex, Mucoid Enteropathy:

Young rabbits often show a variety of enteric conditions ranging from constipation to a profuse watery diarrhea. Although no one has clearly defined one cause for all these enteric disorders, a number of distinct infectious entities may be involved. These include a number of agents such as Bacillus piliformis, Salmonella spp., colibacillosis, intestinal colibacillosis, intestinal coccidiosis, and a clostridial enterotoxemia.

This disease complex usually occurs in 7-10 week old rabbits which have just been weaned and placed on a new diet. Clinical signs include loss of appetite, increased thirst, diarrhea, rapid weight loss, grinding of the teeth, bloated abdomen and a stained anal area with light green to brown feces and mucous. Diagnosis can only be confirmed on necropsy by finding goblet cell hyperplasia. These is no specific treatment due to the uncertainty of an etiology for this disease complex; however, increased roughage in the diet has been shown to reduce the incidence of this disease.

FUNGAL DISEASE

Ringworm is produced by two main groups of fungi, Trichophyton spp. and Microsporum spp.. These fungi cause patchy areas of dry, crusty, thickened skin. Lesions are usually found on the nose, ears, eyelids and feet of nursing young where it can produce devastating results. Single lesions are also found in the adult, but they are much less severe. Diagnosis is made by finding fungi in skin and hair scrapings and by culture. Treatment involves a thorough cleaning of all nest boxes and nesting materials with a strong detergent, hot water, chlorine bleach and/or a lye water rinse, and cleaning up the nursing mothers with soap and water. In severe cases, all diseased animals should be given griseofulvin at a dosage of 25 mg/kg/day for two weeks or until lesions disappear. There is also a topical ointment or solution available for use directly on lesions in milder cases (miconazole) (Table 2). Other topical treatments include daily cleansing with 0.1% chlorhexidine diacetate (Nolvasan®, Fort Dodge, Iowa, USA); and nystatin. This disease is also transmissible to humans and should be reported to the local healthy authority.

PARASITIC DISEASES

External Parasites:

The most common external parasite of rabbits is Psoroptes cuniculi, which produces ear mange or canker. The mites live in the ear canal producing a brown, waxy material, which soon forms a dark crusty covering composed of mites, dried blood, and cellular debris. If not treated the ear may lose its vascular supply and actually slough or fall off the head. Contrary to some beliefs, external ear mites do not produce a middle ear infection, but excessive scratching can cause secondary bacterial infection.

Treatment for ear mites can involve commercial products such as Mitox® (Norden Laboratories, P.O. Box 80909, Lincoln, Nebraska, USA, 68501), or a 0.25% suspension of lindane in mineral oil. Approximately 6-8 drops of this suspension or Mitox should be placed in the ear and massaged every third day for two weeks. Mineral oil by itself will tend to suffocate the mites and can provide a fairly effective treatment. The use of small amounts of kerosene in plant oil such as red palm oil has been used to keep ear mite problems in check (see Table 3). The most effective treatment is to use Ivermectin® injectable (Merck and Company, Inc., Rahway, New Jersey, USA, 07065). This product has not been approved for rabbits in the United States, but is very effective (Table 3).

Wild rabbits may have infestation with ticks of the genus Haemaphysalis. Most housing facilities are sufficient to break the life cycle and so they are usually not a problem in domesticated rabbits. Another parasite of the wild rabbit is the grub worm (larvae) of the fly, Cuterebra spp.. Grub worms hatch from fly eggs and burrow into the skin to form warbles. These larvae, sometimes three quarters of an inch long, can be removed by enlarging the opening in the skin and pulling them out with tweezers. The wound is then cleaned with an antiseptic.

Internal Parasites:

Without a doubt, the most common parasitic disease in rabbits is coccidiosis. The most severe intestinal forms are Eimeria magna and Eimeria irresidua. These parasites produce diarrhea, poor appetite, weight loss and sometimes death. Diagnosis is dependent on finding the oocysts in the feces on microscopic examination. Control of intestinal coccidiosis is directly related to management practices which minimize fecal contamination of feed, water and hutch floors. Most breeders use wire bottom floors and it has been shown that agitation of the floor on a daily basis will increase the flow of feces through the floor and thus greatly reduce the incidence of coccidiosis. Cages may also be soaked in a 10% ammonia solution, which will kill the oocysts (Table 3). In many cases this is all that is necessary to break the life cycle of the parasite; however, if further treatment is required amprolium, sulfaquinoxaline, sulfamerazine, and sulfadimethoxide have been proven to be effective (Table 3).

One species of coccidia, Eimeria stiedae, is known to attack the liver and is therefore very pathogenic. This parasite enters the body through the intestine, but it soon migrates to the bile ducts of the liver. On necropsy, the liver will have characteristic white, circular nodules on its serosal (outer) surface. Control measures are the same and are the most effective means of regulating this parasite. Medications will provide temporary relief as before, but the parasites will soon reinfect the colony if the life cycle is not broken.

Tapeworms:

The rabbit tapeworm, Cittotaenia variabilis, is a flat, ribbon-shaped parasite which usually causes little or no problem in low numbers. However, if excessive numbers are reached they can produce diarrhea and weight loss. Good sanitation will usually control this problem. Medications used have included praziquante and niclosamide (Table 3).

The dog releases eggs of the tapeworm Taenia piriformis in its feces and is thus a source of infection for the rabbit. These larvae enter the digestive tract, migrate through the liver leaving white scars and eventually form small fluid-filled cysts (cysticerci) in the abdominal cavity. There is no treatment for the larval form and so control is accomplished through good management.

Roundworms:

Three roundworms are seen in the rabbit. Obeliscoides spp. is a very slender, reddish colored stomach worm and Trichostrongylus spp. is a similar parasite in the small intestine. These two parasites are usually not a problem and if they are present they can be easily controlled with good sanitation. The third roundworm is Passalurus ambiguus, a pinworm living primarily in the cecum. These parasites are glistening white, about one half inch long, and rarely cause any disease problems. Again, management practices will usually break the life cycle. All roundworms can be effectively treated with piperazine citrate at a dose of 100 mg/100 m of drinking water for one day. This treatment must be repeated in 14-21 days. Other drugs which have been used include phenothiazine and fenbendazole (Table 3).

Table 6.3
SELECTED DRUGS AND DOSAGES

DRUGS	DOSAGES	SOURCES
Antimicrobals		
Cephalothin	12.5 mg/kg BW IM qid for 6-8d	
Chloramphenicol palmitate	50 mg/kg BW PO tid for 7-10d	Chloromycetin , Parke-Davis Oral Suspension, Fort Dodge
Chloramphenicol sodium succinate	30 mg/kg BW IM sid for 5-7d	Chloromycetin® , Park-Davis
Gentamycin	4 mg/kg BW IM sid for 5-7d	Garacin® , Schering-Plough Garasol® , Schering-Plough Gentocin® , Schering-Plough Tech America
Penicillin G, Benzathine or Penicillin G, Procaine	60,000 IU/kg BW IM bid for 7-10d	PFI-Pen G™, Pfizer Crysticillin® 300 A.S., Solvay Go-Dry, GC Hanford Pen-Aqueous, Tech America
Penicillin G, Benzathine and Procaine combination	42,000-84,000 IU/kg BW SQ once weekly for 3-7 treatments or 0.45 mL/kg BW SQ every other day for 5-7 treatments	Flo-cillin® , Fort Dodge Pen BP-48, Pfizer Crystiben® , Solvay Dual-Pen, Tech America GC Hanford
Tetracycline or Oxytetracycline	15-50 mg/kg BW PO bid or 400-1000 mg/L drinking water	Terramycin, Pfizer Panmycin Aquadrops® Upjohn Polyotic Powder, American Cyanamid Tetracycline HCL, Richlyn Tetracycline HCL, Premo Tetracycline Powder 324, Tech America Oxytet Soluble, I.D. Russel
Sulfadimethoxine	75-100 mg/kg BW PO sid for 7d	Albon , Roche Bactrovet , Pitman-Moore

Table 6.3
SELECTED DRUGS AND DOSAGES

DRUGS	DOSAGES	SOURCES
Sulfamethazine <u>or</u> Sulfamerizine	100 mg/kg BW, PO, sid <u>or</u> 625 mg/L drinking water Example 1: 5 mL of a 12.5% solution in 1.0L water Example 2: Dilute 10 mL of a 12.5% solution with 40 m water. Give 10 mL PO bid to an adult (5 kg) rabbit for 3-5d, skip 3 days, repeat	Spanbolet II®, Smith Kline Beecham Sulfa-S™, Smith Kline Beecham Sulfatech™ SR, Tech America Sulmet®, American Cyanamid Sustain III™, Sanofi Veta-Meth, Vet-A-Mix
Furazolidone	5 g/L drinking water for 7-10d <u>or</u> mix 1:40 in feed (50 g/ton)	Furall, Farnam Furoxone®, Smith Kline Beecham NF-180 Suspension, Hess and Clark
Nitrofurazone	100 mg/L drinking water for long term treatment <u>or</u> 2.4 g/L drinking water for 5-7d	Furacin®, Smith Kline Beecham Enteritis Formula, Veterinary Laboratories Nitrofurazone Solution, Veterinary Laboratories Fura-vet, Biomed Laboratories NFZ, Hess and Clark Nitrofurazone Soluble Powder, Tech America
Sulfaquinoxaline	250 mg/kg feed (250 g/1000kg) <u>or</u> 0.05% in drinking water for 10d	Bova-Cox®, Sanofi Sulfa-Nox Concentrate or liquid, Purina Mills Sulquin 6-50 or Soluble powder, Solvay

Table 6.3
SELECTED DRUGS AND DOSAGES

DRUGS	DOSAGES	SOURCES
Antifungal Agents		
Griseofulvin	25 mg/kg BW PO sid for 14d or 250 mg/kg BW PO once every 10d until lesions disappear Example: Mix one 250 mg tablet with 10 mL water and give 1.0 mL/kg BW PO sid	Fulvicin®, Schering-Plough
Miconazole nitrate	Apply lotion or cream to affected areas once daily until lesions disappear	Conofite®, Pitman-Moore
Parasiticides		
Fenbendazole	50 ppm (50 mg/kg) food for 5d	Panacur®, Hoechst-Roussel
Niclosamide	150 mg/kg BW PO; repeat in 14-21d	Niclocide®, Miles Pharm. Division
Phenthiazine	1 g/50 g feed (may need to add sweeteners to the feed)	Phenothiazine granules, Fort Dodge
Piperazine	Add 1 g/L to drinking water (10.1% solution) for 1 day; repeat in 14-21 days or 25 mg/kg BW PO sid for 5d	Pipa-Tabs, Vet-A-Mix Pulvex Caps, Zema Liquid Wormer, Purina Mills Wazine 34, Solvay Sergeants Worm-Away, Conagra Kennel Wormer, Happy Jack

Table 6.3
SELECTED DRUGS AND DOSAGES

DRUGS	DOSAGES	SOURCES
Amprolium	0.15% solution (1.5 mg/mL) in drinking water for 2 days. Then 0.007% solution (0.07 mg/mL) for 4 days. Repeat in 2 weeks Example 1: 310 mL of 9.6% solution (Corid®) in 20 L water; then 15 mL in 20 L water Example 2: 150 g of 20% powder (Corid®) in 20 L water; then 7.5 g in 20 L water	Corid 9.6% solution <u>or</u> 20% powder, MSD Agret Amprovine 25%, MSD Agret
Metronidazole	0.025% solution in drinking water Example: 45 g active ingredient in 190 L water	Flagyl® , GD Serle & Co. Metric® 21, Fielding Protostat® , Ortho Pharmaceuticals Metromidazol tablets, Pharmatair; Par; Barr; Danbury; Lederle; Martec; Squibb
Benzyl benzoate	Use as a 2-5% dust in talcum powder	
Ivermectin	0.1-0.2 mL SQ of a 1:100 dilution in sterile propylene glycol (Ivomec®) or sterile water (Eqvalan®)-1	Eqvalen® , Ivermectin® MSD Agvet

Table 6.3
SELECTED DRUGS AND DOSAGES

DRUGS	DOSAGES	SOURCES
Ammonia	10% solution used on caging following a thorough cleaning with hot water to kill coccidial occysts	Generic
Mitox™	6-8 drops in each ear every 3d for 2 weeks. Repeat as necessary	Mitox™ liquid, Smith Kline Beecham
Kerosene/ Palm oil	Mix 10 drops Kerosene in 20 mL (0.25/) red palm oil; use several drops in each ear	
Oxytocin	0.2-3.0 u/kg BW SQ	Tech America, Anthony, Forbes
Calcium gluconate	5-10 mL of a 10% solution PO <u>or</u> 3-5 mL IV	Tech America, Cal-Dextro® , Fort Dodge
Sodium Hypochlorite	1.5 mL of a 5-25% solution (bleach) in 7.6 L drinking water (200 mg/mL) or free chlorine to chlorinate drinking water	Clorox® ; Clorox Company Many generics
Hydrochloric Acid	3.0 mL of a 1.0 N hydrochloric acid solution per L water for the acidification (pH:2.4) of drinking water	Generic

KEY:		
	BW = body weight	d = days
	g = grams	kg =kilograms
	IM = intramuscular injection	
	IV = intravenous injection	IU = international units
	L = Liters	mg = milligrams
	PO = per os, by mouth, oral dose	
	SQ = subcutaneous injection	u =units
	bid = twice daily	gid = four times daily
	sid = once daily	tid = three times daily

DRUG SOURCES: The following is a list of drug companies from the United States that produce the drugs in Table 3. We have tried to include telephone numbers in addition to addresses so that you may be able to contact the company directly to check the availability of a certain drug. Most of the companies are veterinary pharmaceutical companies. Manufacturers of generic drugs (*) usually produce less expensive drugs. Check with these companies about the availability of any other drugs that may have lost their patent since the printing of this text.

1. American Cyanamid Company, One cyanamid Plaza, Wayne, NJ, USA, 07470, (201) 831-2569.
2. Anthony Products Company, 5600 Peck Road, Arcadin, CA, USA, 91006, (800) 423-7153 *.
3. Barr Laboratories, Inc., 2 Quaker Road, P.O. Box 3-2900, Pomona, NY, USA, 10970, (914) 362-1100 *.
4. Biomed Laboratories, 438 West Arrow Highway, Unit 30, San Dimas, CA, USA, 91773 *.
5. The Chlorox Company, P.O. Box 24305, Oakland, CA, USA, 9462, (800) 292-2200.
6. Conagra Pet Products, 3902 Leavenworth, Omaha, NE, USA, 68105 *.
7. Danbury Pharmacal, Inc., 131 West Street, P.O. Box 296, Danbury, CT, USA, 06810, (203) 744-7200 *.
8. Farnam Companies, Inc., 301 West Osborn, Phoenix, AZ, USA, 85013 *.
9. The Fielding Company, 2384 Centerline Industrial Drive, St. Louis, MO, USA, 63146, (314) 567-5462.
10. Forbes Laboratories Inc., 800 Fifth St., Fort Dodge, IA 50501, (515) 955-4600 *.
11. Fort Dodge Laboratories Inc., 800 Fifth Street, Fort Dodge, IA, USA, 50501, (515) 955-4600 *.

12. G.C. Hanford Manufacturing Co., 1017 Syracuse, NY, USA, 13201 *.
13. Happy Jack Inc., Box 475, Snowhill, NC, USA, 28580 *.
14. Hess and Clark, Inc., Seventh & Orange Streets., Ashland, OH, USA, 44805 *.
15. Hoechst-Roussel Agri-Vet Company, Somerville, NJ, USA, 08876, (800) 247-4838.
16. Lederle Laboratories, Division of American Cyanamid Co., One Cyanamid Plaza, Wayne, NJ, USA, 07470, (914) 735-2825 *.
17. Martec Pharmaceutical Inc., 106 West 11th, Suite 2100, Kansas City, MO, USA, 64105, (800) 822-6782 *.
18. Miles Inc., Consumer Healthcare Division, 1127 Myrtle Street, Elkhart, IN 46515, (219) 264-8955.
19. MSD Agret, Division of Merck & Co. Inc., P.O. Box 2000, Rahway, NJ, USA, 07065, (201) 855-3800.
20. Ortho Pharmaceutical Corporation, Raritan, NJ, 08869, USA, (201) 218-6000.
21. Par Pharmaceutical Inc., One Ram Ridge Road, Spring Valley, NY, USA, 10977, (800) 828-9393/(914) 425-7100 *.
22. Parke-Davis, Division of Warner-Lambert Company, 201 Tabor Road, Morris Plains, NJ, USA, 07950, (201) 540-2000.
23. Pfizer, Inc., Animal Health Division, 235 E. 42nd St., New York, NY, USA, 10017, (212) 573-7646 *.
24. Pharmafair, Inc., 110 Kennedy Drive, Hauppauge, NY, USA, 11788, (800) 227-1427, (516) 231-0707 *.
25. Pitman-Moore Inc., 421 E. Hawley St., Mundelein, IL, USA, 60060, (708) 949-3300.
26. Premo Pharmaceutical Labs, 111 Leunig St., S. Hackensack, NJ, USA, 07606 *.
27. Purina Mills, Inc., P.O. Box 66812, St. Louis, MO, USA, 63166 *.
28. Richlyn Laboratories, Inc., Castor of Kensington Avenues, Philadelphia, PA, USA, 19124 *.
29. I.D. Russell Co. Laboratories, 1301 Iowa Street, Longmont, CO, USA, 80501 *.
30. Roche Animal Health and Nutrition, Hoffman-LaRoche Inc., 340 Kingsland Street, Nutley, NJ, USA, 07110, (201) 235-5000.
31. Sanofi Animal Health, Inc., Veterinary Products, 7101 College Blvd., Overland Park, KS, USA, 66210, (913) 451-3431/(800) 538-2382.
32. Schering-Plough Animal Health, P.O. Box 529, Kenilworth, NJ, USA, 07033, (201) 709-2500.

33. G.D. Serle & Co., Box 5110, Chicago, IL, USA, 60680.
34. Smith Kline Beecham Animal Health, 812 Springdale Drive, Exton, PA, USA, 19341.
35. Solvay Animal Health, Inc., 1201 Northland Drive, Mendota Heights, MN, USA, 55120-1139, (612) 681-9555 *.
36. E.R. Squibb & Sons, Inc., P.O. Box 4000, Princeton, NJ, USA, 08543-4000, (609) 921-4000 *.
37. Tech America Veterinary Products, Fermenta Animal Health, 10150 N. Executive Hills Blvd., P.O. Box 901350, Kansas City, 2 MO, USA, 64190, (816) 891-5500 *.
38. The Upjohn Company, Animal Health Division, 7000 Portage Road, Kalamazoo, MI, USA, 49001, (616) 323-4000.
39. Vet-A-Mix, Inc., 604 W. Thomas Ave., Shenandoah, IA, USA, 51601, (712) 246-4000 *.
40. Zema Corp., P.O. Box 12803, Research Triangle Park, Durham, NC, USA, 27709 *.
41. Veterinary Laboratories, 12304 Santa Fe Drive, Lenexa, KS, USA, 66215 *.

Section 7

Sanitation And Salughtering

SANITATION PROGRAM

An essential part of disease prevention and reproduction is a good daily sanitation program. Environmental factors such as the proper type of hutch and shelter construction, and a routine program for cleaning and maintenance are particularly important. Ideally, cages should be cleaned daily with a wire brush to remove loose hair and manure. Following a thorough scrubbing, the cages should be brushed or sprayed with a chlorine bleach solution (30 mL of chlorine bleach per liter of water). The combination of scrubbing and bleach solution is an effective way to control coccidiosis and enteritis. The build-up of calcium carbonate deposits may be reduced by rinsing these areas with a weak acid solution, such as vinegar. "Izal" is an inexpensive disinfectant available in Western Africa in 125 m tins which dilutes to make 20L. This may be used to clean hutches, furniture, feeders, and water bowls. Be sure to clean water and food bowls daily.

Avoid overcrowding of the rabbits. Build the hutches in an area which has good drainage, fresh clean water and adequate sunlight. Avoid damp, poorly ventilated areas as this will only reduce the animals resistance to disease and infection.

Because feed may act as a mechanical carrier for infection, be sure it is stored in a dry, vermin-proof container, such as a metal can with a sealed lid on it. Also, properly designed feed hoppers are necessary for any good sanitary program. Avoid scattered feed in the rabbitry, as this will only serve to attract insects, mice, rats and birds. Of course, all of these are potential carriers of disease.

Once a month, soiled hutches or cages should be removed from the rabbitry and thoroughly disinfected. In addition, pens or buildings housing infected animals should also be disinfected. One agent, which acts as both cleaner and disinfectant, is lye or caustic soda. One 13-ounce (370 mL) can of lye can be dissolved in 57 liters of water or one pound (454 g) makes about 76 liters of disinfectant. Care must be taken to avoid getting this lye solution on painted surfaces, clothing, aluminum or the skin of rabbits and humans, as it may cause severe burns.

Sunlight is also an effective disinfectant; however, the exposed equipment must be thoroughly cleaned of all debris prior to exposure as the sun has very little penetrating power. When possible, a cement slab, exposed to the direct rays of the sun, provides a good place to disinfect movable equipment. Other methods which can be used include a variety of commercial chemical disinfectants, live steam, and dry heat such as that produced by the direct flame of fire. Care must be taken when using any of these agents to avoid human injury.

In addition to maintaining the suggested sanitation and disinfection practices, the daily inspection and handling of sick animals is important. A good preventive medicine program would include:

1) Purchase only healthy animals. Never bring diseased animals into your rabbitry. Quarantine all new animals for thirty days before introducing them into your colony.
2) Isolate all sick animals in a completely separate area. If the animal cannot be readily cured, it is probably best to destroy these animals and bury or burn the carcasses.
3) Be sure to care for healthy animals prior to servicing the disease animals. This will prevent carrying organisms into your healthy colony.
4) Mark all pens which contained diseased animals. Be sure to thoroughly clean and disinfect these cages with a 2% lye solution (one pound, 454 g, of lye in 20 liters of water) before placing new rabbits in them.

SLAUGHTERING AND DRESSING

The rabbit that is ready for slaughter can be quickly and humanely killed by delivering a quick, sharp blow to the base of the skull. An alternate method is to dislocate the neck by holding the rabbit up by the hind feet, then pulling the head in a quick upward snap.

The carcass then can be suspended by the hind legs by slitting the skin between the tendon and the tibia. Hooks, nails or rope can be used to tie the legs and then suspend the rabbit from an overhead beam, tree limb, etc. (Figure 35).

Next, cut the neck to allow free bleeding or remove the head. The tail is removed and a cut is made from one hock to the other. The skin is cut completely around the hock area. Next cut through the skin around the anus and genital area extending this cut to the cut running from leg to leg (Figure 36).

The edge of the skin is now separated from the carcass and, starting at the legs, pull inside-out over the body. The front legs are removed at the carpus (wrist) before the skin from the body is pulled over this area. Likewise, the head should be removed from the body if not already done. Fat should be left on the carcass. The entire hide is then pulled off the rabbit and remains flesh side out (this is referred to as a cased skin). The cased skin can be stretched over a wood or wire form for drying (Figure 37).

Now remove the internal organs. First make a cut through the muscle on the midline of the abdomen from below the anus to the end of the sternum (breastbone). Then cut the cartilage alongside the breastbone, opening the chest area. Pull the intestines out of the body cavity, using your knife to cut any tough attachments.

Remove the liver, heart and lungs. The gallbladder will lie to one side of the center of the liver. It is necessary to remove the gallbladder from the liver if the liver is to be saved and eaten, as fluid leakage from the gallbladder can give the liver an unpleasant taste.

After removing the internal organs, the carcass should be rinsed in cold water for a short period not to exceed 15 minutes. This helps remove any hair or blood on the meat and cleanse the carcass.

Next, the carcass can be cut into sections (Figure 38). Remove the front legs by cutting from the underneath side of the legs, up through the shoulder. Likewise, cut the hind legs free at the hip. Now cut down to the vertebrae just behind the last ribs. Then bend the spine back to locate a space between the vertebrae and cut there to separate the carcass in two. Cut the ribs from the backbone (vertebrae) on each side to make two rib sections. The pelvic area can, likewise, be split in two or left as one large piece.

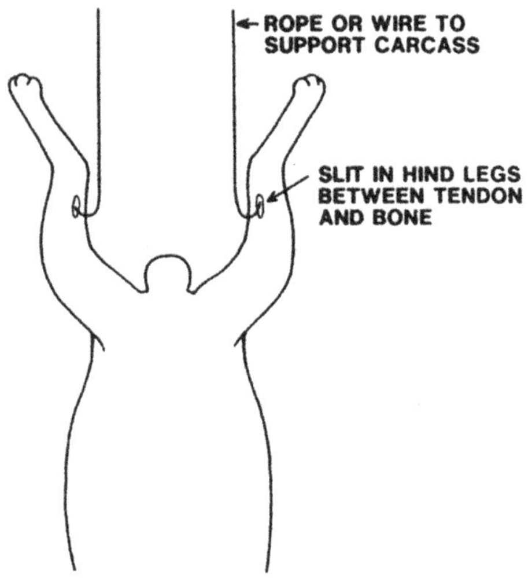

Figure 35: Suspending carcass by hind legs.

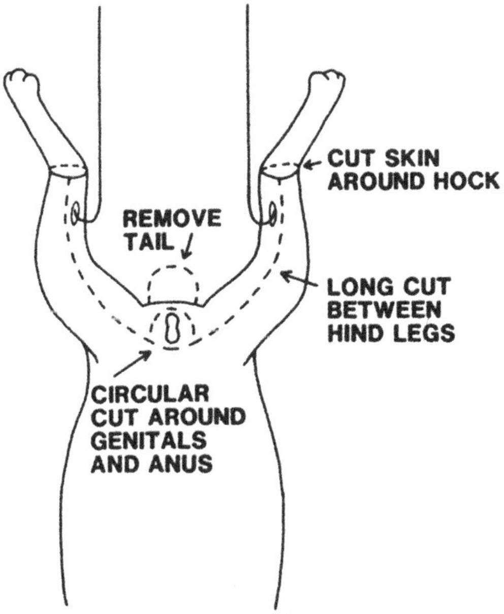

Figure 36: Dotted lines show proper cuts for hide removal.

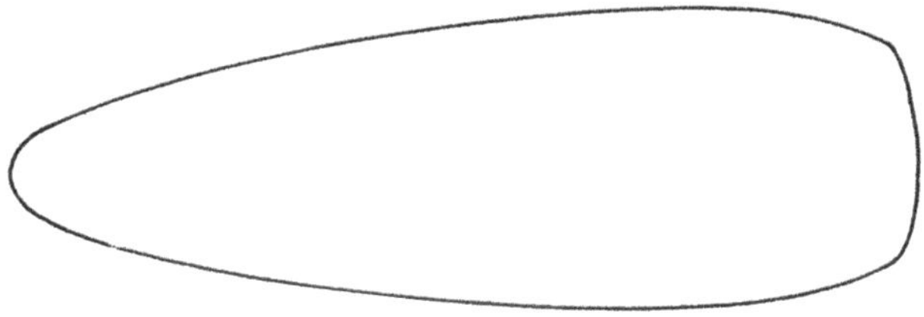

Figure 37: General shape for wood or wire form for drying cased pelt.

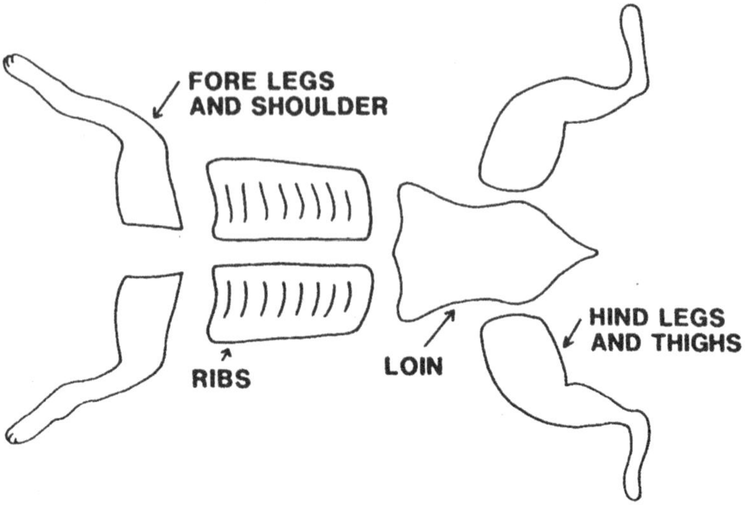

Figure 38: Method of dressing rabbits for marketing.

Section 8

Rabbit Production In Various Parts Of The World

RABBIT PRODUCTION IN VARIOUS PARTS OF THE WORLD

INDONESIA (PACIFIC ISLANDS)

In certain areas of Indonesia, villagers may only have the opportunity to eat meat 2-3 times annually. Because of this protein shortage, the government is encouraging village rabbit production. Three or four does and a single buck provide enough fryers to supply meat for the village weekly. They are fed predominantly forage which is collected by the villagers. These foodstuffs include the leaves of tree legumes (Leucaena and Gliricidia spp.) which are high in protein, elephant grass, sweet potato vines, vegetable leaves (cabbage, carrot tops, etc.), roadside grasses, and cassava leaves. The tubers of cassava are also utilized (for people) as they are high in starch. Feeding leucaena leaves as the sole forage reportedly causes hair loss and poor growth. Setaria grass also has been found to result in poor weight gains.

Even heavy breeds such as Flemish Giants are housed on split bamboo floors with no sore hock problems. The major disease seen is scabies (mange), caused by the mite Sarcoptes scabiei. The small local breeds, probably developed by Dutch settlers, are utilized as they are well adapted to the environment, and are also cross-bred with imported breeds such as New Zealand Whites, Californians, and Flemish Giants.

CHINA (EASTERN ASIA)

China dominates the world market in the production of Angora rabbit wool, but has a large sustenance farmer population as well. Rabbits are raised in a wide variety of situations, ranging from brick and concrete rabbitries, brick hutches with bamboo floors, to all bamboo cages. Feedstuffs vary, but many easily obtained wilted forages are available, including black locust (both wet and dried), water lettuce (Pistia stratiotes), water peanut (Alternanthera philoxeroides), sweet potato leaves, alfalfa, and Chinese vetch.

NEPAL (CENTRAL ASIA)

In Nepal, rabbits are often kept in wooden or bamboo hutches. The report given to the authors states that rabbits do well simply on dry grass. If there is grass left over from day to day that is not wet or soiled, then this is taken as an indication that the grass is sufficient as a diet in and of itself. Otherwise, grains and "trash" vegetables may be fed. Salt, either mixed with grains, as a block, or in a shallow bowl, is given to rabbits free choice.

CAMEROON (WEST AFRICA)

Rabbits are often kept in hutches constructed of native raffia palm placed on poultry wire under eucalyptus pole barns with raffia palm siding and thatched, zinc, or aluminum roofs. They are fed maize bran daily with supplements including plantain, banana, palm nuts, sweet potatoes, and avocados which are locally available and at times quite inexpensive or surplused. A local plant, African iodine (Aspelia latiola), has been used to non-specifically but successfully treat enteric disease. Forages fed ad libitum (free choice) include desmodium (Desmodium distortum, D. intortum), molasses grass (Meiinis minutiflora), elephant grass (Pennisetum pupureum), Stylosanthes guianensis, Guantemala grass (Tripsacum laxum), brachiaria (Brachiaria ruziziensis), and blackjack (Bidens pilosa).

Does are generally bred back only after weaning the kits at 56 days of age. Thus four litters per year are expected of them, producing about 20 fryers annually. More intensive breeding would likely require a higher plane of nutrition than is possible. During the dry season when foodstuffs are rare, many producers stop kit production to reduce the amount of feed necessary. Guatemala grass, which can provide green forage even during these months, is planted anticipating the dry season. Otherwise, increased kitchen scraps or purchased food is necessary.

An interesting technique used here (and other places) integrates animal production by placing rabbit facilities directly over ponds. Manure and waste food fall directly into the pond to serve as food and fertilizer for fish in the pond. Plants that grow easily around the rim of the pond are fed to the rabbits as well.

HAITI (CARIBBEAN ISLANDS)

Producers in Haiti have been very successful despite the warmer-than-ideal temperatures. Feedstuffs are provided entirely by local crops and by-products and include a mash of brewers' grain, wheat shorts, molasses, water, and rice bran. Other ingredients in this formula include sea salt, limestone, an imported mold inhibitor (GV 11) and copper sulfate. This is fed at about 85 g (3 oz) per day per animal. Forages fed free-choice include bermuda grass, kudzu, corn leaves, cassava leaves, sweet potato vines, wilted water hyacinth greens and flowers, sugar cane, and local greens. It has been noted, however, to take up to 18-24 weeks to get a fryer to 1.8 kg (4 lbs) live weight on this diet.

THE AUTHORS

W. SHELDON BIVIN

Director, Dept. of Vivarial Science and Research,
Tulane School of Medicine
1430 Tulane Avenue, New Orleans, LA. 70112

B.S., 1962, Kansas State University
D.V.M., 1964, Kansas State University
Ph.D., 1971, University of Missouri

WILLIAM W. KING
Clinical Veterinarian and Graduate Student
Division of Laboratory Animal Medicine
School of Veterinary Medicine
Louisiana State University
Baton Rouge, Louisiana 70803
Telephone: 504-346-3145

B.S., Biology, 1987, Rhodes College
D.V.M., 1991, Louisiana State University

Dr. King is a clinical veterinarian at the Louisiana State University Division of Laboratory Animal Medicine and graduate student in the Department of Veterinary Pathology. Although his research involves clinical pathology and diagnostics in non-mammalian species, he is responsible for the daily veterinary care of thousands of laboratory animals.

REFERENCES

Ashbrook, F.G. and C.E. Kellogg. Rabbit Production. Farmers Bulletin No. 1730, U.S. Dept. of Agriculture, 1934, 1939, 1950.

The Biology of the Laboratory Rabbit. Edited by Steven H. Weisbroth, Donald E. Flatt and Alan L. Kraus, Academic Press, 1974.

Cassady, R.B., P.B. Sawin, and J. Van Dam. Commercial Rabbit Raising. Agriculture Handbook 309, U.S. Dept. of Agriculture, 1966, pp.69

Cheeke, P.R. 1983. Rabbit production in Indonesia. Journal of Applied Rabbit Research. 6:3, 80-89.

Cheeke, P.R. and N.M. Patton. 1987. Rabbit production and research in the People's Republic of China. Journal of Applied Rabbit Research. 10:2, 68-75.

Domestic Rabbits: Diseases and Parasites, Agricultural Handbook No. 490, U.S. Dept. of Agriculture, 1976.

Harris, D.J. Construction of Quonset-Style Rabbit Cages. J. Applied Rabbit Research, 142-174.

Kanable, A. Raising Rabbits. Rodale Press, Emmaus, Pa., 1977, pp. 191.

Kentor, W.E. 1990. Rabbit raising in Haiti. Journal of Applied Rabbit Research. 13:2, 69-70.

Lukefahr, S.D. and M. Goldman. 1987. Cameroon, West Africa: economic feasibility of rabbit farming under intensive and subsistence management systems of production. Journal of Applied Rabbit Research. 10:1, 20-25.

Lukefahr, S.D. 1985. A technical assessment of production and economic aspects of small-scale rabbit farming in Cameron. Journal of Applied Rabbit Research. 8:3, 126-135.

Milne, G. 1982. A first-hand look at Chinese rabbit production. Journal of Applied Rabbit Research. 5:2, 54-60.

Self-Teaching Manual on Backyard Rabbit Rearing. Food and Agriculture Organization of the United Nations Regional office for Latin America and the Caribbean Santiago-Chile, July, 1986.

Templeton, G.S., F.G. Ashbrook and C. Kellogg. Rabbit Raising. Conservation Bulletin No. 25, U.S. Fish and Wildlife Service, 1994.

Christian Veterinary Mission (Publisher of this book)

Our vision is to see

Christ's love expressed through veterinary medicine.

Our mission is to

challenge, empower and facilitate veterinarians to serve through their profession, living out their Christian faith.

CVM also provides education and encouragement for those who desire to minister through service, prayer, relationship building, and modeling Christ's love.

About CVM

Christian Veterinary Mission (CVM) is a registered non-profit Christian Service Organization 501(c)(3) based in Seattle, Washington, U.S.A.

CVM was founded in 1976 by Dr. Leroy Dorminy who came to realize the impact that veterinarians could have by integrating their faith with their practice, both locally and around the world. In 2008, CVM had nearly 30 veterinary professionals serving full-time internationally and over 200 veterinary professionals and student volunteers serve on short-term cross-cultural mission trips annually. CVM sponsors fellowship & prayer breakfasts at over 20 U.S. veterinary meetings each year and reaches out to veterinary students through Christian Veterinary Fellow-ship (CVF) groups in every veterinary school in the U.S. by encouraging them in spiritual growth and professional development.

There are over 3,500 veterinarians affiliated with CVM in the U.S. CVM also partners with organizations and networks in other countries that are focused on empowering Christian veterinarians. CVM has a volunteer advisory board of veterinarians who guide its vision, mission, and programming.

CVM books and the free International Animal Health Newsletter were written with small farmers, veterinarians, and agricultural development workers in mind. Our desire is that they would help individuals and groups develop an appropriate livestock program to meet community needs. CVM's Endowment Fund was started in the early years of the organization's life. The fund provides for meaningful programs that could not be funded by the regular budgeting process.

NOTES

Made in the USA
Monee, IL
04 November 2026

67379266R00066